# Robotics at Home with Raspberry Pi Pico

Build autonomous robots with the versatile low-cost Raspberry Pi Pico controller and Python

**Danny Staple**

BIRMINGHAM—MUMBAI

# Robotics at Home with Raspberry Pi Pico

**Group Product Manager**: Rahul Nair
**Publishing Product Manager**: Rahul Nair
**Content Development Editor**: Sujata Tripathi
**Technical Editor**: Rajat Sharma
**Copy Editor**: Safis Editing
**Project Coordinator**: Sean Lobo
**Proofreader**: Safis Editing
**Indexer**: Pratik Shirodkar
**Production Designer**: Aparna Bhagat
**Senior Marketing Coordinator**: Nimisha Dua
**Marketing Coordinator**: Gaurav Christian

First published: March 2023

Production reference: 1170223

Published by Packt Publishing Ltd.
Livery Place
35 Livery Street
Birmingham
B3 2PB, UK.

ISBN 978-1-80324-607-9

www.packtpub.com

*To my amazing wife, Carol, who has supported all my robotics experiments with love, inspiration, and tolerance of a living room filled with robots. To my children, Helena and Jonathan, for all their ideas and enthusiasm and for encouraging me to take a break and play with them sometimes!*

*– Danny Staple*

# Contributors

## About the author

**Danny Staple** is a robot builder and programmer. He has been a professional software engineer since 2000, uses Python professionally, and regularly contributes to open source projects.

Danny has been building robots at home since 2004 and has a cupboard full of projects, including robots with wheels, cameras, tank tracks, legs, and arms, made from plastic, cardboard, metal, kits, lunchboxes, and modified toys.

Danny authored *Learn Robotics Programming*, published in 2021 by Packt Publishing, and has written magazine articles for *The MagPi*. He runs the robotics YouTube channel **Orionrobots** and brings his robots to events such as Pi Wars and Arduino Day. Danny also mentors at CoderDojo KU, where he shows kids how to program in Python and has run Lego robotics clubs.

*I would like to thank the Pi Wars and Adafruit communities for answering my tricky questions, Mike Moncrieffe for checking diagrams for me, and my review team for the great feedback throughout this book.*

# About the reviewer

**Leo White** is a professional software engineer and a graduate of the University of Kent. His interests include electronics, 3D printing, and robotics. He first started programming on the Commodore 64 and later wrote several applications for the Acorn Archimedes. He currently programs set-top boxes for his day job. Utilizing the Raspberry Pi as a foundation, he has mechanized children's toys and driven robot arms, blogging about his experiences and processes along the way. He has given presentations at Raspberry Jams and entered a variety of robots in the Pi Wars competition.

# Table of Contents

# Part 1: The Basics – Preparing for Robotics with Raspberry Pi Pico

## 1

# 5

# Driving Motors with Raspberry Pi Pico    101

# Part 2: Interfacing Raspberry Pi Pico with Simple Sensors and Outputs

# 6

# Measuring Movement with Encoders on Raspberry Pi Pico    121

# 9

# Teleoperating a Raspberry Pi Pico Robot with Bluetooth LE  197

# Part 3: Adding More Robotic Behaviors to Raspberry Pi Pico

# 10

# Using the PID Algorithm to Follow Walls  221

# 11

# Controlling Motion with Encoders on Raspberry Pi Pico  249

# 12

# Detecting Orientation with an IMU on Raspberry Pi Pico  271

# 13

# Determining Position Using Monte Carlo Localization          291

# 14

# Continuing Your Journey – Your Next Robot          339

# Preface

Robotics is an emerging field with applications in every walk of life. Robotics, and the associated technology, appear to be confined to the well-equipped laboratories of universities and high-tech companies. However, many of the aspects of robotics – building them and programming them – can be learned and practiced in your own home.

The main areas of robotics are as follows:

- Structure – the design and building of a mechanical platform
- Electronics – sensors, motors, and control circuits
- Software – the code for libraries, sensor interactions, and behaviors

This book aims to cover a little of each area, looking at basic CAD design, part fabrication, and assembly of hardware. It introduces some starting digital electronics, such as connections and data buses. It aims to dig a little deeper into the sensors and the code needed to make interesting behaviors using them.

There are robotics books that offer a theoretical robotics introduction; however, the aim of this book is to take you on a journey of practice, fun, and experimentation. This book provides step-by-step applied explanations and images to aid understanding.

Building your own robots in your home is a great way to learn technology skills. This is an experience of technology that replaces impenetrable magic with real-world experience and confidence to build more – anyone with practice can become a robotics wizard too.

## Who this book is for

The book is intended for those who would like a practical and step-by-step hands-on introduction to designing, building, and programming robots, using the popular Python programming language. It is also for those who would like to gain an introduction to 3D CAD, robotics sensors, robotics hardware, and robotics behaviors that make use of the sensors and hardware.

This book will be valuable to makers, learners, and developers who want to build robots in their homes or workshops. The book does not require a specialist workshop, and any skills and tools needed will be explained throughout the book.

Those who have written a little code before will find this book useful. You do not need to have any experience with electronics or making things, but you can expect to gain initial experiences while practicing the techniques in this book.

We expect you to have a keen interest in learning more and a little fearlessness in trying robotics experiments. Practical application of the examples within is essential. Getting the most out of this book means being willing to make a real robot and test it.

# What this book covers

*Chapter 1*, *Planning a Robot with Raspberry Pi Pico*, introduces Raspberry Pi Pico in relation to other robotics main controllers. It shows the advantages of the CircuitPython programming environment and takes you through making an overview plan for a robot build built around Pico. The chapter provides a robot hardware shopping list for the first half of the book, discussing the parts and trade-offs in choosing them.

*Chapter 2*, *Preparing Raspberry Pi Pico*, takes you through getting CircuitPython onto Pico, then taking your first steps in writing code with it. It will also cover soldering headers onto Raspberry Pi Pico so it can connect to robot parts.

*Chapter 3*, *Designing a Robot Chassis in FreeCAD*, introduces FreeCAD while turning the overview plan into 3D CAD designs. It shows you how to make drawings from the design for building the robot parts.

*Chapter 4*, *Building a Robot around Pico*, shows how you can use CAD drawings with hand tools to craft robot parts by cutting and drilling sheet plastic. It guides you in assembling the parts then wiring and connecting the electronics. This chapter is where the robot is first powered on!

*Chapter 5*, *Driving Motors with Raspberry Pi Pico*, introduces you to controlling motors with CircuitPython and Raspberry Pi Pico, showing how motors can be used to make line motions and turns and how speed can be controlled. The chapter then shows you how to pull these together into programmed motion sequences.

*Chapter 6*, *Measuring Movement with Encoders on Raspberry Pi Pico*, introduces the first robotic sensor in the book with wheel encoders, showing you how to detect wheel movement in code. The chapter covers the Raspberry Pi Pico PIO peripheral as a powerful way to manage these sensors.

*Chapter 7*, *Planning and Shopping for More Devices*, prepares you for the next section of the book with distance sensors, Bluetooth LE, and an **inertial measurement unit** (**IMU**), with further advice on choosing the devices and how they will be attached. The chapter provides a shopping list for the latter part of the book. You will revisit FreeCAD part design to make sensor mounts, and then use tools to cut them.

*Chapter 8*, *Sensing Distances to Detect Objects with Pico*, takes you through attaching and wiring two distance sensors into the robot. The chapter provides information on I2C communication and then shows you how to program the robot to communicate with the sensors. You will then build code for the robot to autonomously avoid walls.

*Chapter 9, Teleoperating Raspberry Pi Pico Robot with Bluetooth LE*, makes a comparison of wireless connection options, showing why Bluetooth LE was a suitable design choice. You will connect a Bluetooth LE module to the robot, then extend existing code to output sensor data through this connection, and display the output on a smartphone. You will also see how to drive the robot from a smartphone.

*Chapter 10, Using the PID Algorithm to Follow Walls*, provides an introduction to the PID algorithm, a fundamental building block for sensor/output control behaviors in robotics. We build a wall-following demonstration using a distance sensor, then show you how to tune the PID with smartphone plots via Bluetooth LE.

*Chapter 11, Controlling Motion with Encoders on Raspberry Pi Pico*, revisits encoders, showing you how to convert their output into units understandable by humans. You will learn how to combine these sensors with the PID algorithm to control motor speeds and drive in a straight line. You will then program the robot to drive a specified distance in a straight line at a specified speed.

*Chapter 12, Detecting Orientation with an IMU on Raspberry Pi Pico*, introduces the IMU, a sensor that lets you determine the orientation of the robot. The chapter provides a guide on connecting the sensor and calibrating it. You will use the IMU with the PID algorithm for a behavior that makes a robot always face north. Finally, the chapter shows you how to program the robot to make a specified turn using the IMU.

*Chapter 13, Determining Location with Monte Carlo*, will show you how to program a robot to determine where it is likely to be in an arena. You'll use plans in the chapter to build a foam board arena and model this arena in code. You are shown how to visualize this space on a computer using Bluetooth LE with Matplotlib. You will then learn about moving robot poses based on sensor input. The chapter shows how multiple robot behaviors can cooperate in the same application. You will be introduced to using probability algorithms in robot motion, making predictions, and refining them.

*Chapter 14, Continuing Your Journey – Your Next Robot*, provides a summary of the topics learned in the book, with information on digging deeper into each of them. The chapter provides ideas and research areas for you to extend all the aspects of the robot, and then further suggestions to build more ambitious robots and grow your skills. The chapter also recommends robotics communities you could participate in.

## To get the most out of this book

You will need to have knowledge of a few Python basics, such as variables, looping, conditionals, and functions. A well-lit and ventilated desk space is recommended for the robot-building aspects of the book. Access to hand tools will help, although you will be shown which tools to shop for. The robot code examples have been tested on CircuitPython 7.2.0 on Raspberry Pi Pico but should work with later versions. The computer code examples were tested on Python 3.9.

| Software/hardware covered in the book | Operating system requirements |
|---|---|
| Thonny > 3.3 or Mu Editor > 1.1 | macOS, Linux, or Windows |
| Python 3.7 or later | macOS, Linux, or Windows |
| Matplotlib 3.6.1 or later | macOS, Linux, or Windows |
| NumPy 1.23.4 or later | macOS, Linux, or Windows |
| Bleak (Python BLE library) 0.19.0 or above | macOS, Linux, or Windows |
| Free USB port | macOS, Linux, or Windows |
| Smartphone/tablet with Bluetooth LE (Bluetooth > 4.0) | iOS or Android |
| Adafruit Bluefruit LE Connect > 3.3.2 | iOS or Android |
| Bluetooth LE-enabled laptop (or BLE dongle) | macOS, Linux, or Windows |
| FreeCAD | macOS, Linux, or Windows |
| Raspberry Pi Pico | |
| CircuitPython > 7.2.0 | Raspberry Pi Pico |

Thonny comes with a built-in Python 3.x installation. The **Tools | Open System shell** menu can be used to install packages in Thonny's Python.

**If you are using the digital version of this book, we advise you to type the code yourself or access the code from the book's GitHub repository (a link is available in the next section). Doing so will help you avoid any potential errors related to the copying and pasting of code.**

Help for this book can be found by:

- Raising a bug on the book's GitHub repository at https://github.com/ PacktPublishing/Robotics-at-Home-with-Raspberry-Pi-Pico

- Asking via Discord at https://discord.gg/2VHYY3FkXV

## Download the example code files

You can download the example code files for this book from GitHub at https://github.com/ PacktPublishing/Robotics-at-Home-with-Raspberry-Pi-Pico. If there's an update to the code, it will be updated in the GitHub repository.

We also have other code bundles from our rich catalog of books and videos available at https://github.com/PacktPublishing/. Check them out!

# Download the color images

We also provide a PDF file that has color images of the screenshots and diagrams used in this book. You can download it here: https://packt.link/7x3ku.

# Conventions used

There are a number of text conventions used throughout this book.

`Code in text`: Indicates code words in text, database table names, folder names, filenames, file extensions, pathnames, dummy URLs, user input, and Twitter handles. Here is an example: "To run this code, be sure to send the `pio_encoders.py` library, the updated `robot.py` file, and then `measure_fixed_time.py`."

A block of code is set as follows:

```
import time
import board
import digitalio

led = digitalio.DigitalInOut(board.LED)
led.direction = digitalio.Direction.OUTPUT

while True:
    led.value = True
    time.sleep(0.5)
    led.value = False
    time.sleep(0.5)
```

When we wish to draw your attention to a particular part of a code block, the relevant lines or items are set in bold:

```
>>> print("Hello, world!")
Hello, World!
>>>
```

Any command-line input or output is written as follows:

```
code.py output:
4443 4522
```

**Bold**: Indicates a new term, an important word, or words that you see onscreen. For instance, words in menus or dialog boxes appear in **bold**. Here is an example: "Launch Mu Editor, and when it is running, click on the **Mode** button. From this, select **CircuitPython**."

> **Tips or important notes**
> Appear like this.

# Get in touch

Feedback from our readers is always welcome.

**General feedback**: If you have questions about any aspect of this book, email us at customercare@packtpub.com and mention the book title in the subject of your message.

**Errata**: Although we have taken every care to ensure the accuracy of our content, mistakes do happen. If you have found a mistake in this book, we would be grateful if you would report this to us. Please visit www.packtpub.com/support/errata and fill in the form.

**Piracy**: If you come across any illegal copies of our works in any form on the internet, we would be grateful if you would provide us with the location address or website name. Please contact us at copyright@packt.com with a link to the material.

**If you are interested in becoming an author**: If there is a topic that you have expertise in and you are interested in either writing or contributing to a book, please visit authors.packtpub.com.

## Share Your Thoughts

Once you've read *Robotics at Home with Raspberry Pi Pico*, we'd love to hear your thoughts! Scan the QR code below to go straight to the Amazon review page for this book and share your feedback.

https://packt.link/r/1803246073

Your review is important to us and the tech community and will help us make sure we're delivering excellent quality content.

# Download a free PDF copy of this book

Thanks for purchasing this book!

Do you like to read on the go but are unable to carry your print books everywhere? Is your eBook purchase not compatible with the device of your choice?

Don't worry, now with every Packt book you get a DRM-free PDF version of that book at no cost.

Read anywhere, any place, on any device. Search, copy, and paste code from your favorite technical books directly into your application.

The perks don't stop there, you can get exclusive access to discounts, newsletters, and great free content in your inbox daily

Follow these simple steps to get the benefits:

1.  Scan the QR code or visit the link below

https://packt.link/free-ebook/9781803246079

2.  Submit your proof of purchase
3.  That's it! We'll send your free PDF and other benefits to your email directly

# Part 1: The Basics – Preparing for Robotics with Raspberry Pi Pico

In this part, you will take your first steps in learning about Raspberry Pi Pico, then plan and build a robot around it, and get the initial robot code to make the robot move.

This part contains the following chapters:

# 1

# Planning a Robot with Raspberry Pi Pico

When you plan, you create the best chance for a mission's success. We want to build robots in an achievable way. Let's start with a plan in mind! We'll use this plan to explore why **Raspberry Pi Pico** is a great fit for this and make a shopping list!

In this chapter, you'll learn about Raspberry Pi Pico's capabilities. You'll discover **CircuitPython** and understand why it is a great language for Raspberry Pi Pico. Additionally, we'll plan a robot design and understand the trade-offs to make choices about the robot early in the project. We'll check that our robot fits together, working out the parts and tools you'll need with suggestions on how to get them.

At the end of the chapter, you'll have both a plan and parts arriving so that you are ready to build a robot. Additionally, you'll have a starting process for making other robots and setting yourself up for success with them.

In this chapter, we'll cover the following main topics:

- What is Raspberry Pi Pico, and why is it suitable for robotics?
- What is CircuitPython?
- Planning a Raspberry Pi Pico robot
- Test fitting a Raspberry Pi Pico robot
- A recommended shopping list for robot basics

## Technical requirements

We'll go into the necessary hardware and shopping list as we progress further in this chapter. So, in this section, we'll just focus on what you will need physically and on your computer to get started.

You will require the following:

- Some thin cardboard
- A ruler, pencil, and scissors
- A good web browser with internet access

# What is Raspberry Pi Pico, and why is it suitable for robotics?

At the heart of every robot is a **controller**. Usually, this is a computing device that is responsible for running the code for the robot to perform its tasks and behaviors. Choosing a controller is a key choice in robot design. You can either come from the *I have this controller, what can I do with it?* perspective or the *which controllers have the capabilities I'll want for a particular robot?* perspective.

In this section, we'll take a closer look at what Raspberry Pi Pico offers as a controller and the trade-offs it's made. We'll explore why it is good for robotics and why it could be part of a larger, more interesting system, too.

Additionally, we'll delve into the details of its interfaces and how they'll be useful to us.

## A microcontroller that runs Python

Let's start by taking a look at Raspberry Pi Pico, and discover what it has. The following photograph shows Raspberry Pi Pico:

Figure 1.1 – Raspberry Pi Pico

Raspberry Pi Pico, as shown in *Figure 1.1*, is an *RP2040* microcontroller on a Raspberry Pi-designed board. This **microcontroller** is a small computing device that has been designed to interface closely with hardware. It has a USB connection on the right-hand side for power or programming on a computer. The LED is useful for debugging. Also, there are many **input/output** (**IO**) pins around the edges to connect things. It is with these IO pins that the magic happens when it comes to controlling robots!

Controllers use IO pins to write and read from attached hardware. They can group pins into buses (which we'll cover in more detail later) to exchange data with other devices. Additionally, they can create waveforms on outputs for controlling motors and LEDs.

This sounds a lot like the other **Raspberry Pi** models. However, this is a different class of computer. **Raspberry Pi Pico** has more in common with an **Arduino** board. Let's take a closer look at what that difference means with the following diagram:

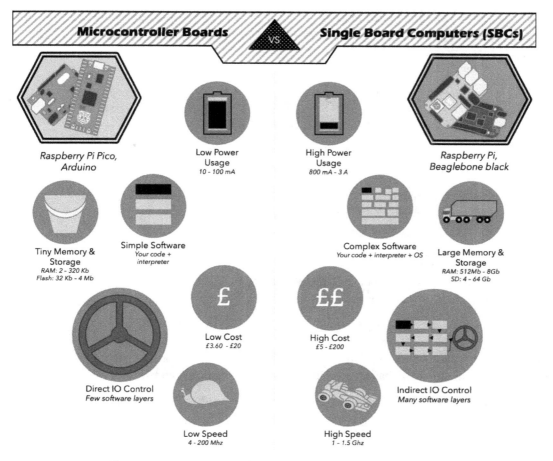

Figure 1.2 – Microcontroller boards versus single-board computers

*Figure 1.2* shows that while microcontroller boards such as Raspberry Pi Pico and Arduino might look similar to **single-board computers** (**SBCs**) such as Raspberry Pi 4 or BeagleBone, they have different key areas. For instance, they differ in storage, CPU speed, cost size, the complexity of software, and how closely your software runs to the hardware.

While Raspberry Pi Pico is brilliantly suited to controlling hardware, such as robots, it isn't as suited to high-memory or CPU tasks such as AI or visual recognition. There's a kind of robot system known as **horse-and-rider**, which combines an SBC (for example, Raspberry Pi 4) for complex processing with a microcontroller (for example, Pico) for controlling hardware.

The low complexity means that code on a microcontroller has nearly no boot time, which means your code doesn't have to coexist with other software in an operating system. Take a look at the following block diagram:

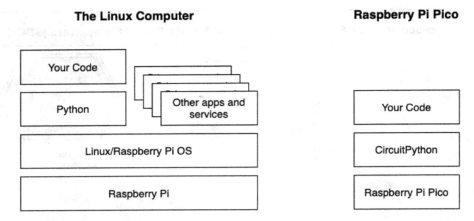

Figure 1.3 – Running your code on Raspberry Pi versus Pico

This preceding diagram represents the software architecture on Raspberry Pi versus Raspberry Pi Pico. It shows how a Linux computer, such as Raspberry Pi, has additional layers of software along with competing apps running alongside your code.

In addition to this, controllers have **interrupts**. They can notify the code that something has changed, such as the state of an IO pin. You'll find this on the other Raspberry Pi models, but they are controlled by that pesky operating system again. In Pico and other microcontrollers, you get more control over what happens or when something changes on an IO pin, allowing responsive code with predictable timing.

So, how does Raspberry Pi Pico compare with the Arduino Uno? The following table shows details from their specifications and datasheets:

| Specification | Raspberry Pi Pico | Arduino Uno |
| --- | --- | --- |
| Digital IO pins | 26 | 14 |
| Analog IO pins | 4 | 6 |
| Processor | Dual Core RP2040 at 133 MHz | Single-core ATmega328 at 16 MHz |
| Flash (storing code) | 2 MB | 32 KB (plus 1 KB EPROM) |
| RAM (running memory) | 264 KB | 2 KB |

Table 1.1 – Comparing the Pico with the Arduino Uno

The preceding table shows that Raspberry Pi Pico has a faster multicore processor, along with more storage and digital IO pins. Additionally, Raspberry Pi Pico has a unique **Programmable IO (PIO)** system for extreme flexibility in organizing data to and from these pins. Official Pico boards are also cheaper than official Arduino boards.

Another place that Raspberry Pi Pico compares favorably with Arduino is in its use of Python (CircuitPython or MicroPython). Many microcontrollers, such as Arduino, require C/C++ to program, which can be difficult for beginners. Python is easier to understand, allows for complex and interesting data structures, and has access to many libraries of code, too.

In short, the key features of Raspberry Pi Pico are as follows:

- A microcontroller—this offers low power and is small compared with SBCs.
- It has responsible and flexible IO options.
- It is low cost compared to many microcontroller boards and most SBCs.
- It is programmable in Python.

A number of the features I attribute to Raspberry Pi Pico are due to the *RP2040*—the chip that powers Pico and is available in forms other than Raspberry Pi Pico.

IO flexibility is Raspberry Pi Pico's most interesting feature, so let's take a look at that next.

## Raspberry Pi Pico's interfaces for sensors and devices

Raspberry Pi Pico has many interfaces for connecting to hardware, along with its unique PIO system. In this section, we'll look at each type of interface.

A **digital IO pin** is the basic IO system for Raspberry Pi Pico. An output can be on or off, which is great for turning LEDs on or off, but you are unable to control their brightness. Similarly, an input can also detect on or off states. Raspberry Pi Pico has 26 of these pins.

**Pulse-Width Modulation (PWM)** is a waveform for controlling outputs such as LEDs and motors—including DC motors, stepper motors, and servo motors. PWM pins output square wave pulses, with a changing (modulating) on-off ratio (pulse widths). Changing pulse width results in changes to the brightness of an LED, the speed of a motor, or a servo motor's position. Raspberry Pi Pico has 16 PWM channels, making it capable of controlling many such devices at once. These PWM pins still require a power control device to drive the motors.

**Analog** input pins detect levels of voltage between **ground (GND)** and 3.3V. This is good for interfacing with simple sensors, such as light sensors, joysticks, slider/knob controls, temperature sensors, and measuring currents (using a bit of additional circuitry). Raspberry Pi Pico has three of these inputs.

A **universal asynchronous receiver-transmitter** (**UART**) controls a serial port. It can send streams of data to and from devices using two pins: a **TX transmit** pin and an **RX receive** pin. With this, it is capable of sending/receiving data that is more complicated than just a varying level. Raspberry Pi Pico has two independent UART interfaces.

Pico has two **Serial Peripheral Interface** (**SPI**) bus controllers. SPI uses four pins, as shown in the following diagram:

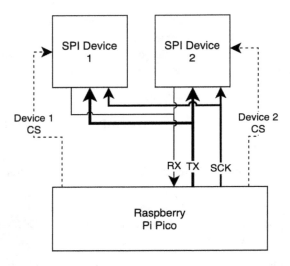

Figure 1.4 – Raspberry Pi Pico SPI bus usage

The preceding diagram shows Raspberry Pi Pico using an SPI bus to connect to two devices—for example, displays or sensors. The bus has **transmit** (**TX**), also known as **Controller Out/ Peripheral In** (**COPI**) or **Microcontroller Out/Sensor In** (**MOSI**) for transmitting data from the controller, **receive** (**RX**) also known as **Controller In/ Peripheral Out** (**CIPO**) or **Microcontroller In/Sensor Out** (**MISO**) for receiving data back to the controller, **SCK** (a clock for timing the signal), and **Chip Select** (**CSEL/CS**) a chip selection pin for each peripheral. SPI uses chip selections to enable communication with multiple devices, as shown by the dashed lines of **Device 1 CS** and **Device 2 CS**. See `https://makezine.com/article/maker-news/mosi-miso-and-140-years-of-wrong/` for details on the current SPI acronyms.

The **Inter-Integrated Circuit (I2C)** is a data bus designed for communicating between integrated circuits such as sensors, memory devices, and output devices. An I2C bus has a data pin (which is often called *SDA – Serial Data*) and a clock pin (which is often called *SCL – Serial Clock*) keeping things synchronized. Multiple devices share an I2C bus by sending/receiving data with addresses, such as those in the following diagram:

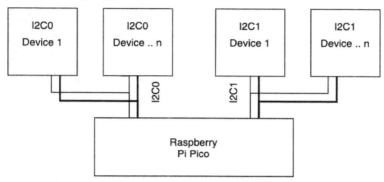

Figure 1.5 – I2C buses on Raspberry Pi Pico

*Figure 1.5* shows Pico and then some child peripherals connected via two independent I2C buses, assignable to different pin configurations, with some devices having the same address but different I2C connections. Additionally, I2C can address registers (such as memory locations) within devices. We'll use I2C later to communicate with sensors.

Finally, Raspberry Pi Pico has PIO. PIO is a feature that is unique to Pico. PIO consists of two blocks with four *state machines*. Each can run simple code independently of the main CPU and control one or more pins to send data to or from them. A single-state machine can control all the pins if that was useful for the code. Additionally, each state machine comes with buffers to hold data until it can be transferred. The following is an example block diagram of the PIO system:

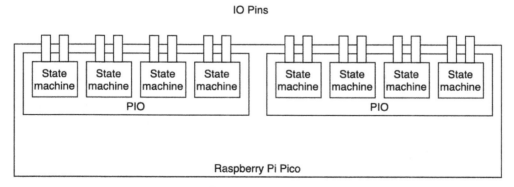

Figure 1.6 – The Raspberry Pi Pico PIO system

The preceding diagram shows two PIO devices inside the Pico. Each has code storage memory, so you can have two independent functions. In each PIO device, there are state machines that can independently run the code from that local memory.

Since PIO state machines run independently, and their instructions are about shifting data to/from pins, they can create interfaces for many kinds of hardware. For example, is there a weird protocol device? Use PIO. Do you need rapid counting independent of the main CPU? Use PIO. People have made **Video Graphics Array** (**VGA**) outputs with PIO, so it's capable of fast and complex data handling. Additionally, you can also get interrupts from PIOs to tell you when something has happened.

That was quite a lot of IO systems. Let's summarize them in a table, as follows:

| Name | Direction | Number usable | Typical uses |
| --- | --- | --- | --- |
| Digital IO pin | In/out | 26 | Turning LEDs on/off. Detecting button presses. |
| PWM | Out | 16 | Controlling varying outputs, such as motor speeds. |
| Analog input pin | In | 3 | Reading varying voltages on sensors. |
| UART | In/out | 2 | Sending/receiving complex data packets. |
| SPI | In/out | 2 | Sending/receiving complex data packets. |
| I2C | In/out | 2 | Sending/receiving complex data packets. |
| PIO | In/out | 2 x 4 | Complex control of digital pins—including data packets, pulse counting, and shifting in clocked data. Protocols are not covered by the other data buses. Independent timed control of output pins. |

Table 1.2 – The Raspberry Pi Pico IO systems

These protocols share pins, so using an I2C bus consumes 2 pins from the 26-pin pool.

Now that we've had a tour of Raspberry Pi Pico's features and interfaces, let's take a look at how we'll program it in this book, using **CircuitPython**.

## What is CircuitPython?

Many microcontrollers require C/C++ or Assembler to program—for example, the popular Arduino ecosystem. However, in robotics, Python is rapidly becoming a de facto language. It is used for AI and data science and is great for rapidly trying out new ideas. Let's examine why it is handy and, specifically, why I've chosen **CircuitPython** for this book.

Python does not require a compile step. Getting you quick feedback on your code and Python's **read-eval-print loop** (**REPL**) allow you to start typing and experimenting with code instantly. The REPL allows you to see what works before using ideas in code that you'll keep. Here's a REPL session with CircuitPython:

```
Adafruit CircuitPython 6.2.0 on 2021-04-05; Raspberry Pi Pico
with rp2040
>>> print("Hello, world!")
Hello, world!
```

The preceding session shows a print running in a REPL on Raspberry Pi Pico. We'll explore how to use the REPL for some Pico experiments. It even comes with built-in assistance; however, on Pico, not all of the help is left in, for size reasons.

Python has other things that help, such as being able to directly return multiple values from a function. Python has function calls and classes like C++, but functions can be used as data, and references to them can be stored in variables. Additionally, Python has functional programming elements that allow programmers to chain tools together for processing streams of data.

Python uses exceptions to handle errors, allowing you to choose how to respond to them or observe their output, leading you directly to a problem.

MicroPython is the original port of the Python language to run on small memory devices such as microcontrollers. It has a community working on it, and CircuitPython builds on it.

In CircuitPython, Raspberry Pi Pico mounts as a USB storage device, so you can copy your code and the libraries your code uses, directly onto the Pico. This makes composing code from multiple libraries or using third parties simple. Copying code over with the correct name is enough to run that code when Raspberry Pi Pico is powered up again.

CircuitPython has a huge library of device support for Neopixel LEDs, Bluetooth, many sensors, displays, and other devices. This library not only works with Pico but runs across many CircuitPython controllers, so familiarity with these library components will be useful when you are working with other controllers.

Now that we've chosen a language and the controller that we will build robots with in this book, it's time to start planning a robot!

# Planning a Raspberry Pi Pico robot

We've been fact-finding for our robot-building mission. Before we start our robot-building journey, we'll make a rough plan of what we want to do, then refine it. We'll make important decisions, which we can examine further as we start to build the robot.

## An overview of robot planning

When planning the robot, there are several things we need to consider:

- What do we want this robot to do? What is it for?
- What style of robot is suitable?
- What kinds of sensors or outputs will we need?
- What rough shape and size will it have?

Once we've answered these questions, we can make further decisions about what we build. These don't require much detail. Robotics is full of interesting diversions, making it tempting to jump between ideas. By having a constrained plan and working to it, you can keep your pace on getting a robot built, saving distractions and cool ideas for the next robot or three!

### What do we want this robot to do? What is it for?

Will the robot solve a problem, clean your kitchen, explore a space, deliver packages, impress guests at a conference, or compete in a robot competition?

The robot we'll build in this book has several purposes:

- Exploring Raspberry Pi Pico and its capabilities
- Trying out sensors
- Writing algorithms guided a little by challenges in robot competitions
- Navigating a known space
- Building a custom chassis, adaptable for future ideas
- Keeping it simple enough to get started

With these goals in mind, we can look at the specific details.

### What style of robot is suitable?

There are many robot styles. We should choose one, probably the simplest possible for our goal. Take a look at the following diagram for a selection of different robot styles:

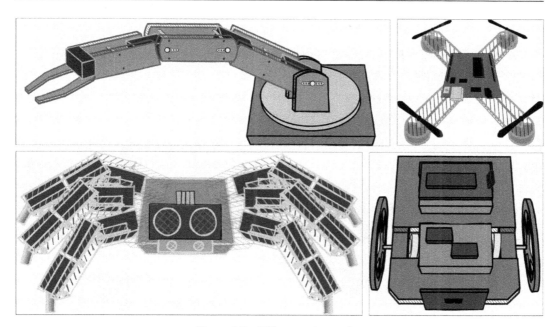

Figure 1.7 – Different robot styles

The first robot style is a robot arm used in industry. These are fascinating and fun to build. However, they do not satisfy our goals of building a robot chassis to try out sensors.

The next panel shows a quadcopter drone—an **unmanned aerial vehicle (UAV)**. These are complicated to build and program, so they do not meet our goal of keeping it simple.

The third panel shows a walking robot—a hexapod. These require controlling many servo motors. Their power usage and complexity make them an unsuitable but exciting option for a follow-up robot!

The fourth panel shows a wheeled robot. Wheeled robots can be simple **two-wheel-drive (2WD)** robots with a roller, such as this one. 2WD rover platforms such as this satisfy our goals of building a chassis and getting to know sensors and algorithms. They can later be made more interesting, with tracks, mecanum wheels, rocker bogies, or individually steered wheels, allowing them to also meet the adaptable goal.

I recommend that we go ahead with a 2WD rover throughout this book but keep the other variations in mind for further robot builds!

### What kinds of sensors or outputs will we need?

One of our goals is to try out different sensors. A robot made to navigate spaces will influence the sensors we'll use. They all contribute to locating the robot.

Good sensors for this include the following:

- **Distance sensors**: What is in front of the robot? How far are the nearest objects? We might want more than one of these devices.

- **Encoders**: How has the robot moved? How far did it go?

- **Inertial Measurement Unit (IMU)**: Has the robot turned? What is its position relative to north?

Along with these sensors, we can have simple outputs—the motors that we drive wheels with. As a later extension, we could also add Bluetooth to give us some feedback on our robot's status. We don't need to plan all of this yet but leave space for it so that we can extend the robot later.

### What rough shape and size will it have?

Now, we have a firm idea of a 2WD robot. We know it probably needs to support the following:

- Raspberry Pi Pico

- A pair of motors with wheels and a caster

- Many sensors and, later, Bluetooth

- Power for the system, including batteries plus voltage conversion

- A breadboard for wiring all of this together

Although we don't want the robot to be too big, we are going to need some real estate to play with. Let's start with a rough estimate of 150 mm x 200 mm.

So, we've answered some questions about what we want. We will use the next few sections to dive deeper into the planning of this robot, looking at the different aspects of the planning and the choices we'll make. The first of those is to consider trade-offs.

## A note on trade-offs

All designs make trade-offs. The truth is that no design fits all cases, and usually, no design is perfect but will be good enough in the right aspects where it works. We will need to make decisions and read datasheets for parts to also assist us.

One example is size and weight—we already mentioned that we don't want a large robot. After all, we have a limited workbench size, and larger robots require more power, larger motors, and larger batteries. Additionally, we'd need to work with tougher and—likely—harder-to-cut materials. For a different context and goal, perhaps a large, heavier robot would be more suitable. So, the first trade-off is to keep the robot small but not too small—that is, to keep it simple.

We've suggested Raspberry Pi Pico, and the trade-offs from Raspberry Pi there, for example lighter weight, reduced cost, and power.

But what of sensor trade-offs? Every sensor has multiple types, which we will dive into in their respective chapters. They differ in price, features, and complexity.

In many aspects, we can trade having more complexity for reduced weight or cost or more features for a higher cost.

## Choosing a robot chassis

We have many options for our 2WD robot chassis. Again, this depends on what we want to learn or achieve. We have stated our goal of building a flexible chassis. Some good options for doing this are as follows:

- Buying a chassis kit
- Adapting a lunchbox or toy
- Doing a scratch build by hand
- 3D printing or laser cutting a chassis

Chassis kits are an easy option but have limited flexibility. Many come with motors, wheels, batteries, and even a motor driver designed for a specific main controller. In this way, they can save time and money, allowing you to focus entirely on the code and sensors, but they offer less opportunity to learn design aspects. It's often tricky to find a chassis kit with the right shape and size, and as they get larger, they quickly become more expensive.

You could also adapt a lunchbox into a robot chassis—cutting mounting holes for motors, sensors, boards, and other parts can be a good place to learn design skills. However, you'd need to fit your robot electronics and hardware in a constrained space. Note that the curved sides of lunchboxes can complicate things.

Scratch-building a chassis gives you great flexibility. You can learn how to design in CAD and how to use hand tools. Additionally, you need to make choices about the type and thickness of the material, and in doing so, you'll be able to understand more about making strong robots. You'll learn how to fit sensors and expand your robot if things get a bit tight. This requires more time and patience than the kits, but the rewards are great.

3D printing and laser cutting require precise designs, along with expensive and specialist tools or services. As you dive further into robotics, and progress beyond a simple 2WD robot, creating more interesting shapes and sensor mounts, it is likely to be an important area of exploration. If you are not confident with hand tools, finding a laser-cutting service for the same parts will achieve good results, but it can be costly.

In this book, so that you can get exposure to the design and hand tools while still giving us lots of flexibility, we will take the scratch-building option. We will learn CAD skills that are transferable to 3D printing. We'll learn how to cut and drill parts, looking at some premade parts to save time.

Additionally, we'll size our design at approximately 150 mm x 200 mm and modify this if needed. But what about the motors?

### Choosing motors

This 2WD motor requires two main drive motors. We could consider stepper motors, which move a little each time they are pulsed, although these bring a little extra complexity—perhaps an idea to keep for later. DC motors, which rotate continuously when powered, seem like the right choice. They will need to be geared so that they have enough power to move the robot, without being too quick and hard to control.

We should keep these motors small and at a low voltage. As we are unlikely to want to build an additional gearbox, geared motors are sensible. There are some options here in terms of the size we are working with. First is the yellow *TT* motor with plastic gears— however, these motors are not of great quality and take up a fair amount of space. Another option is to use servomotors that are adapted for continuous rotation—however, these can be a little expensive.

A small, common, high-quality but inexpensive option is N20 or micro-metal gear motors. To save space and effort, there are models of these that have encoders pre-fitted. We can use similarly common plastic brackets to attach them to our robot. That makes them convenient to use, too.

### Robot wheels

For a 2WD robot, there are a few ways in which to lay the wheels out. One possibility is to have two driving wheels with two idler wheels (that is, unpowered). However, those wheels can drag, making it harder to turn the robot. A common way is to have a third wheel as a caster—either a ball that can roll in any direction or a swivel wheel such as a shopping trolley. Because of the size of the robot, a ball caster seems like a good idea.

The wheels themselves should have a hub that is compatible with the motors that we've chosen. A pair of N20 wheels with a diameter of 60-100 mm should be suitable.

So, we have a rough size for our robot, and we know the controllers, motors, and some of the sensors. The next item to choose is the power systems.

## Choosing the power systems

A robot isn't much fun without independent power—by which I mean its own source of power without needing to be plugged into a wall. Usually, this means batteries. It then needs ways to provide power to the control electronics, sensors, external boards, and motors. Take a look at the following diagram for an outline of power distribution in a 2WD robot:

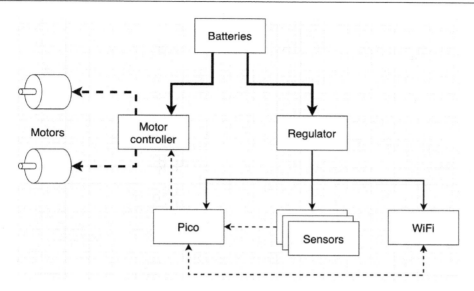

Figure 1.8 – Power distribution in a 2WD robot

In *Figure 1.8*, the thicker lines show raw battery power connections. A 2WD chassis will need to drive at least two motors, which are power-hungry devices that require a battery connection.

This robot needs to power Raspberry Pi Pico and other sensors. Since we intend to add Bluetooth, we should leave power aside for that. So, the other thick line goes to a regulator for these, making more palatable power for these systems—the raw battery voltage would likely destroy them. The thinner solid lines show regulated power.

The Pico will be sending/receiving electronic control signals, designated by the thin dashed lines in the preceding diagram. These also go to the motor controller. The motor controller will provide PWM-controlled power from the batteries to the motor, modulated by the signals the Pico sends to them. The motor power is shown by the thick dashed lines to the motors on the left-hand side.

Here, we have a few considerations to bear in mind. We require an input voltage that is suitable for the motors and to drive a regulator. We need a regulator that can handle the power capacity requirements for the Pico, sensors, and Bluetooth, and we need batteries that can supply enough current to drive them.

### Calculating power requirements

Let's start with what we know—5V is a good output voltage for a regulator, and where needed, the Pico can further regulate down to 3.3V. A regulator for 5V likely requires 7V or more.

> **Important note**
>
> Voltage measures electrical pressure. A current measures how fast electrical energy flows. Combining both of them shows system power usage. A current in amps or milliamps can be used as a stand-in for power in watts when the voltage is known.

Let's look up the specifications for the N20 gear motors. Perform an online search for the N20 motor datasheet. You'll be looking for a PDF document. Usually, these have a picture or diagram of the product, followed by the specification and feature tables. If you look for *Rated Voltage*, they say 6V; however, further down the sheet, there is usually a table relating to the voltage of the motor speed. Based on the motors and regulator basics, an input voltage of 7V-12V would make sense.

Our electronics don't operate on voltage alone and require a current to operate. So, a regulator will need to handle the minimum current requirements. We'll need to look at some datasheets and specifications for the other parts. We will include sensors. For Bluetooth, we will include a low-power **Bluetooth Low Energy** (**BLE**) board using the highest current measurements from `https://learn.adafruit.com/introducing-the-adafruit-bluefruit-le-uart-friend/current-measurements`.

We'll base it on worst-case values so that we can handle them. Let's start by looking at datasheets and gathering numbers into a table, as follows:

| Device | Current requirement |
|---|---|
| Raspberry Pi Pico | Up to 100 mA |
| Adafruit Bluefruit LE UART Friend | Up to 15.2 mA (Bluetooth active) |
| Sensors—distance, IMU, and encoders | Estimate up 50 mA peak |

Table 1.3 – Device power requirements

*Chapter 3* of the datasheet for Raspberry Pi Pico (which can be found at `https://datasheets.raspberrypi.com/pico/pico-datasheet.pdf`) shows the electrical specification, with peak currents at a little under 92 mA (milliamps—a measure of current). We'll round this up to 100 mA as a margin.

The Adafruit Bluetooth board uses only 15.2 mA when fully active, but we can round it up to 20 mA to be generous. The sensors need maybe 50 mA of extra room to accommodate them.

We can add these estimates together to suggest a minimum current specification. Based on these datasheets and estimates, any regulator capable of over 400 mA will be plenty.

## Choosing a regulator

Motor control and power supplies can be totally separate concerns, giving great flexibility, but this can take up more space. There are boards designed for Raspberry Pi Pico that allow you to control motors while supplying power to Raspberry Pi Pico. However, will they supply 400 mA?

Kitronik has some neat Raspberry Pi Pico motor boards—a small motor only, along with a larger robotics board with servo motor outputs and a prototyping area. However, the supply for the Pico on the robotics board datasheet has 100 mA for peripherals, which won't work.

Now, we understand that we need a regulator that is capable of outputting 5V, at a minimum of 400 mA. We want something small and simple. A suitable device for this is a **Universal Battery Eliminator Circuit** (**UBEC**). These can handle 3 A. We'd put this through a VSYS pin on the Pico.

## Choosing a motor controller

The suggested motors are small. Motors have a stall current—that is, the power they draw if they are trying to move and block a logical maximum. For the N20s, their datasheet suggests 350 mA at 6V. It might be a little over that, perhaps 550 mA. A motor controller needs to handle a little over this peak per channel—motor controllers that are unable to handle motor load tend to go up in smoke! Take a look at the following two common simple motor controllers:

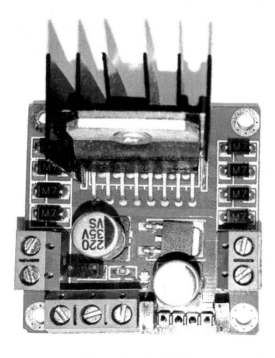

Figure 1.9 – An L298N motor controller next to a TB6612fng

In *Figure 1.9*, on the left-hand side, you can see the very common **L298N**—internet searches suggest this handles 2 A per channel. They are small, inexpensive, and easy to find. However, this is a 5V logic device. Raspberry Pi Pico outputs 3.3V logic, which might not work properly with this controller without logic-level conversion.

On the right-hand side is another good choice, the **Toshiba TB6612fng**. These handle 1.2 A per channel, which is a bit less than shown previously but still good enough. It will handle up to 15V for controlling motors and logic voltages from 3V to 5V, so it won't require level shifters for Raspberry Pi Pico. There is a module from *SparkFun* for these, which will work well with the Pico and is small enough to fit on a breadboard next to it.

Additionally, we should plan to have a power switch on the breadboard from the batteries to the motor controller and regulator. Talking of batteries, we still need to select them.

### Choosing batteries

We also need something to go through the power regulator—batteries.

While we can (and should, at a more advanced stage) consider Lithium-ion batteries, the type used in **remote control** cars, the simplest possible option is AA batteries. For our motor and regulator choice, 8 x AA gives 12V. These are easy to buy and replace but take up a lot of space for their power output. Our motors don't require a lot of current, so they will be good enough.

Now that we've examined our power requirements and some solutions, we'll take a look at which pins are being used on our Raspberry Pi Pico and ensure that our plans for hardware won't conflict.

## Pin usage

Our Raspberry Pi Pico has many IO pins, but we need to consider whether all the items we intend to connect to it will be able to simultaneously connect to it.

Let's re-examine the specifications for the Pico at `https://www.raspberrypi.com/documentation/microcontrollers/raspberry-pi-pico.html`:

- 26 **General Purpose Input/Output (GPIO)** pins
- 2 x UART, 2 x SPI, and 2 x SPI

The motor controller will consume two pins per motor, and we know these motors come with encoders, with a further two pins each. That means, so far, we've used 8 of the 26 IO pins. This should leave us plenty of room for expansion.

Now that we've checked our basic concept, we'll move on to test fitting—an approximation of how we'll build the robot.

# Test fitting a Raspberry Pi Pico robot

Now that we've checked that we can power our system, we need to make sure it is all going to fit on the robot. In this step, you get a rough idea of where things will be, whether your chassis will be big enough, and whether the robot design is likely to work.

The key thing for a test fit is that it is not detailed. Use the simplest method to check whether things will fit, be it sliding around cut-out paper rectangles or using simple software.

Let's make some simple paper or card parts. For this section, you'll require card, a pencil, a ruler, and scissors. Card from a cereal box is great for this, but paper will also do.

For a test fit, rectangles are often good enough. The intention is to determine what will fit inside a space and position things. Detail isn't necessary. For large robots, you might need to make a scale model. As this robot is small, you can make parts at a 1:1 scale. This has an added advantage—if you already have parts in your possession, you get to use them.

## Creating your first test-fit part

You'll need the datasheets for your devices again—this time to start looking at the mechanical sizes of things. For a test fit, you just need to create bounding boxes for items, ensuring there is enough space for them.

Let's start with a breadboard and the Pico. Since the Pico is on the breadboard, you can just model the breadboard size. I recommend a 400-pin breadboard, which is also known as a half-plus. Use a search engine to look for `half plus breadboard dimensions` and click on the images panel. What you are looking for is a flat diagram showing the outside dimensions of the board, such as the following diagram:

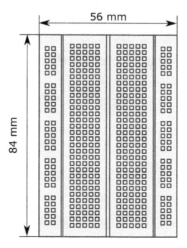

Figure 1.10 – Breadboard dimensions

The preceding diagram shows what to expect with a drawing/picture in terms of dimensions, which, in this case, is a breadboard. It measures 84 mm along the top and 56 mm along the right-hand side. It's important to note here that I'm using millimeters throughout the book, and I will convert from other units if necessary. Stick to one measurement system in a robot design!

We can take the paper or card and draw an 84 mm by 56 mm rectangle in pencil. This does not need to be too accurate—the nearest 5 mm is good enough. The following photograph shows this:

Figure 1.11 – Making a breadboard test-fit part

As the preceding photograph shows, you simply draw it out. Additionally, so that we can identify it later, write *breadboard* on the part, along with its dimensions of *84 x 56*. Keep these handy so that they can be used for reference later.

Then, you can cut this out with scissors. I tend to make a wide rough cut, and then a finer close cut as a second pass for this.

This simple rectangle, with the right measurements, is our first test-fit part. Next, we will need the motor parts.

## Motors

The motors we are using are N20 motors. If we place them on the underside of the robot, as is common with these designs, we still need to consider how their wires come up through the chassis. It is easier to put them on top so that the motor wires can face upward.

We can search the web for N20 motor brackets. Look in the images tab for drawings so that we can get the correct dimensions and add extra space for wiring behind the motor. The following photograph shows me making these test parts:

Figure 1.12 – Making cardboard motor test-fit parts

For this part, as the preceding photograph shows, we want two rectangles of 30 mm by 35 mm. Label them. On one of the longer edges of each part, add an arrow to show that this is where the wheels will go.

We have motors and a breadboard. Next, we need to make stand-in parts for powering them.

## Power systems

The UBEC doesn't take up a lot of space, so we can ignore it. The controller we've chosen will fit on a breadboard with the Pico, so it is already accounted for. We do need to account for the batteries.

We have a couple of variants on an 8 x AA battery holder—the flat kind, which takes up more space but comes with mounting screws, or the 4 x 2 kind. These use vertical space instead. Another way to save space is to put batteries on the underside of the chassis.

At this stage, we will use the flat holder as it is easier. You can look up the size for them and create a labeled rectangle for them. My battery box came out as 93 mm x 57 mm:

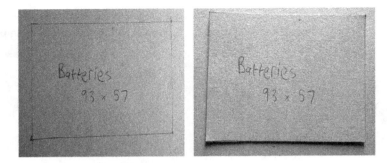

Figure 1.13 – A battery box in cardboard

The preceding photograph shows a battery box created from cardboard. Now, we have a bunch of parts to go on the chassis. Next, we need to represent the chassis itself.

## Creating a rough chassis

We previously suggested that the chassis should be about 150 mm x 200 mm. Create this rectangle in cardboard, as follows:

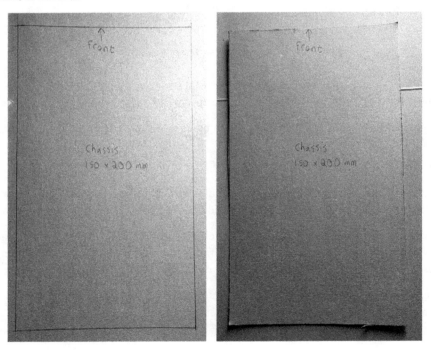

Figure 1.14 – The cardboard chassis

As you can see from the preceding photograph, this is not—at all—intended to be a perfectly neat cut. It is simply meant to be good enough to see where things likely need to go. Give the chassis labels just as we have done so far with the other parts. Additionally, we can label one of the shorter edges of the chassis part as the front.

This is the last item to test fit. Let's start to arrange these parts.

## Arranging the test-fit parts

Now, you should have a set of rectangles representing the different parts. The following photograph shows the parts and how we can arrange them:

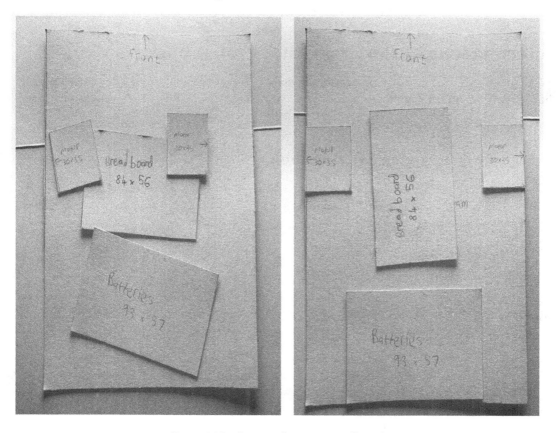

Figure 1.15 – Our test-fit parts in cardboard

The boxes on the left-hand side of *Figure 1.15* are correctly proportioned parts and have been placed in a rough position. However, they aren't properly laid out yet. The right-hand side shows a possible layout.

It'll be easier to fit motors around the breadboard if we rotate it so that it's tall instead of wide. We have the batteries at the back of the robot, in the middle, then we have the breadboard in front of them, along the middle. Notice that there is a gap between the batteries and the breadboard—we don't want any parts to be too close together.

We've put the motors on either side of the breadboard, leaving space at the front to expand our robot design.

We have accounted for the major parts of the robot, including computing, power, and motors. There's also adequate space for expansion. We'll tune this later as we get deeper into the design, but this shows our combination is viable.

Now that we have our robot design, it's time to shop for tools and materials!

# A recommended shopping list for robot basics

We've got a plan—a test fit, a method to make the chassis, decisions about the processing on board, and how we will power the thing. Now, we need to buy the necessary parts and tools to make this happen.

Let's start with the parts.

## Robot parts and where to find them

There are many places to find robot parts, and some of these parts go by different names from different manufacturers. I'll attempt to provide enough information about the parts so that they can be easily found in many countries.

### Part shopping list

We'll start with a part list for the initial robot chassis. For each part, where necessary, we'll show search terms, synonyms, and manufacturer numbers. You can try combinations of these to find parts—I wouldn't suggest using all the terms at once but instead refining them until you get something else. Then, we can discuss places to buy them:

- **Raspberry Pi Pico**: This is the most important part to get you started. Other RP2040-based boards might be suitable; however, with different form factors and pins, you'll need to get creative when it comes to wiring them. Beware of parts with fewer exposed pins, and ensure that they are *RP2040* boards. *Alternative parts*: Headered Pico, Pimoroni Pico LiPo, Adafruit Feather *RP2040*, *SparkFun Pro Micro RP2040*, *RP2040* Plus.

- **A USB micro cable**: You might already have one for your phone, but phones have been coming with USB-C cables for some time. To be specific, I mean a USB-A to USB-Micro cable. However, some laptops have a USB-C port—for those cases, consider a USB-C to USB-Micro cable instead.

- **Headers**: You'll need to solder headers onto Raspberry Pi Pico and the motor controller. Later sensors will require them, too. Search terms for these include *male breakaway strip and single-row PCB through-hole pin headers*. Make sure that you choose 2.54 mm or 0.1-inch pitch headers.

- **A solderless breadboard**: I suggest a mini/half plus 400 tie-point breadboard, self-adhesive. Standard 0.1-inch (or 2.54 mm) pitch spacing, with a separating channel between the columns.

- **Wiring**: You'll need wiring between the components on the breadboard. For this, I suggest precut breadboard jump wire kits. These should be of the U-shape solderless kind—insulated cables with bare ends. They will be in different color-coded lengths and can be bent into shape to fit across breadboard connections.

- **A Drv8833 or TB6612fng motor controller or motor driver**: I recommend the SparkFun or Adafruit models; however, other models will work. Stick to this chip, and ensure it's a module/breakout board, not just the bare chip. It should have a square device with a capacitor on the board, too. I recommend getting a model that has pin functions printed on it. Alternatives are the L9110S, the less efficient L298n boards, or the L293 chip, which may need additional space.

- **A 5V 3A UBEC or a 5V UBEC (DC/DC step-down buck converter)**: Search for ones that can handle a minimum of 3 A. Larger ones are also fine. Look for those with a 5V output. If they allow 6V too, just make sure that you set the jumper to 5V. Look for those with bare input cables and a pin header end. Other buck converter modules may be suitable, but check for 3 A current capacity, and ensure that the output is 5V.

- **8 x AA battery holder or battery compartment**: Look for the flat-style ones with an integrated switch. Some come with round barrel jacks instead of bare ends—in which case, a 2.1-mm jack to screw the terminal block can be used to finish this.

- **A 1N5817 Schottky diode**: These are common in many electronic outlets.

- **Motors**: The N20 micro-metal gear motors with encoders. The 298:1 ratio motors have the right combination of speed and torque. Adafruit has these as ADA4640. Small DC gear motors with encoders can be substituted, but please be aware that the larger motors may require the robot to scale up along with power requirements.

- **Ball caster**: A 16-30-mm caster should be fine. Most models will be suitable. Just remember to use the drawings for their mount holes later.

- **40-90-mm wheels with N20 d-hole or universal hubs**: Make sure they are designed for N20 motors.

- **3 mm or 0.118-inch thick styrene card sheets/plates**: Go for an A4, 200 x 150 mm, or greater size. They should be solid, flat sheets—not foam or hollow, and, ideally, not textured. Possible search terms include Plasticard, styrene sheet, Evergreen, and Plastruct. ABS sheets are suitable, but not acrylic as this can be brittle.

- **Kit for mounting parts**: Possible search terms include standoff and mounting kit. Preferably, they should be in metal, but nylon will do fine. They should be M2, M2.5, or M3 if possible. M2.5 will cover most cases. They should have standoffs, bolts (including machine screws and truss screws), and nuts.

This list of parts and search terms will help you find all the suggested parts to build this robot. Here is a helpful list of places to buy parts from. It isn't exhaustive, and there might be sellers in your country that can help.

For electronics parts, wheels, motors, and modules, the following stores ship worldwide:

- `ThePiHut.com`, `pimoroni.com`, `adafruit.com`, `Robu.in`, and `mouser.com`. For some items, `Pololu.com` and `Sparkfun.com` might carry them.
- Both Adafruit and Pimoroni have web pages that list their various distributors: `https://www.adafruit.com/distributors` and `https://shop.pimoroni.com/pages/worldwide-distributors`. They are a reliable source of reference.
- Online marketplaces such as eBay, AliExpress, and Alibaba can be used—but buyers beware. Sellers on them might not have good support or return policies in place. Parts might also be cheap substitutes, and they could take a long time to deliver.

For the styrene sheet, model supply shops are a good bet. Brands such as Evergreen are widely distributed. Hobbycraft, AliExpress, Alibaba, and Amazon carry these. While there, I suggest getting styrene angle strips and square tubes for later robot enhancements.

## The robot workshop and makerspaces

There is a list of workshop and hand tools associated with this book. You can buy them yourself or use a makerspace. Let's start with the tools that you will need access to.

### List of tools

The tools required for this book are common in many workshops. In addition to this, school and college DT rooms, makerspaces, and many workshop stores will carry them:

- **Plastic cutter**: The kind with changeable blades. We will be cutting through styrene, which can quickly dull a knife. Linoleum flooring cutting blades will also be a useful alternative here.
- A cutting mat to protect your work surfaces.
- You'll need a try square, preferably metal.
- **A ruler**: This should be at least 200 mm long. Since the book is working in metric, I suggest metric markings, too.

- **Sandpaper**: You will need a selection of, at the very least, 400, 600, and 1,000 grit or similar.

- **Soldering iron and stand**: You'll be soldering some parts, so an iron is essential. Do not use a soldering iron without a stand! Many come with them. I recommend a temperature-controlled iron. In addition to this, a brass wool tip cleaner and some solder are required. We'll use lead-free flux core solder wire.

- 10-20-mm hook and loop self-adhesive disks.

- A drill with 2-mm and 3-mm bits. This should be a small drill, preferably cordless. These are small parts, so precision is required more than power.

- You'll need a pencil to make draft lines with. Personally, I like mechanical pencils.

- I recommend safety goggles when you drill or cut. Get a good comfortable pair—cheap uncomfortable ones might end up on your head instead of covering your eyes and won't be protecting you.

- For drilling and cutting, a hobby vise or clamp keeps the part still and your hands safe. A small bench vise is suitable.

- You need a flat work area with good lighting.

- **Screwdrivers**: You will want a screwdriver set. It should have Phillips (PH0 and PH1) and flat-bladed (2 mm and 3 mm) screwdrivers.

- Spanner or wrenches in metric. Precision sets will be useful.

You can purchase these items and tools from electronics, hobby, DIY, and tool stores. AliExpress, Alibaba, eBay, and Amazon will also have them. However, if you do not have all of these tools, it feels like a lot of stuff.

An alternative to buying all these tools is to find a makerspace or hackerspace near you. They will have most, if not all, of these tools. Makerspaces are community-run spaces, have collections of tools, and might even have scrap material of just the right kind of styrene. Additionally, they have other makers, who can lend you a hand and assist you if you run into difficulty with a robot project.

There are makerspaces in most cities globally. They can be easily found on search engines and social media. If there is no makerspace in your area, reaching out via social media to other makers might help you to find a similar community project. There is a global makerspace directory at https://makerspaces.make.co/.

# Summary

In this chapter, you learned about Raspberry Pi Pico, why it's a good controller to build a robot around, and why we will be using CircuitPython to program it. You discovered the planning process of a robot, choosing parts for it, and then test-fitting them to ensure our plan is likely to work. You had a view of datasheets and discovered a little about the size and electrical characteristics of parts. You also had a tour of the parts you'll need to buy and the tools you'll need to work with them. Finally, you were introduced to makerspaces as places to find tools and other makers.

We have a rough robot plan. However, to start building something, we need to take some practical steps to prepare Raspberry Pi Pico for use in one. We'll discover how to do this in the next chapter.

# Exercises

To get you familiar with the content of this chapter, these additional exercises will attempt to test you on what you've learned, and prepare you for later sections:

- Find a datasheet for the Bluefruit LE UART Friend. Find the electrical current used by the device, along with its width and height for fitting it.

- We are going to add such a Bluetooth board to our robot. Use the dimensions from the datasheet to make a part in your test-fit diagram.

- Find a space on the robot for this part that does not overlap with other parts.

- Look on the websites of the previously mentioned stores. See if you can find out where you'd be able to buy this part.

# Further reading

Please refer to the following resources for more information:

- To learn more about CircuitPython, the `https://circuitpython.readthedocs.io/` website is a great resource.

- For a detailed look at Python on microcontrollers, please refer to *MicroPython Cookbook*, by *Marwan Alsabbagh*, *Packt Publishing*. This book has sections on CircuitPython and MicroPython in general.

- *Embedded Systems Architecture*, by *Daniele Lacamera*, *Packt Publishing*, offers an extensive dive into the I2C and SPI bus interfaces.

- Another perspective on makerspaces comes from *Progression of a Maker*, which can be found at `https://hub.packtpub.com/progression-maker/`.

# 2
# Preparing Raspberry Pi Pico

The bare **Raspberry Pi Pico** can run code, but we need some preparatory steps before we can use it. In this chapter, you will see how to get **CircuitPython** up and running and take your first steps in programming Pico. You'll then learn **soldering** so that you can add headers to your Pico—letting you plug it into things.

At the end of the chapter, you'll have Pico ready to run code, with headers ready to add other hardware, and have your laptop or computer ready to send code to Pico. Plus, you'll have some soldering experience if that is new to you.

In this chapter, we will cover the following main topics:

- Getting CircuitPython onto Raspberry Pi Pico
- Preparing the CircuitPython library for Pico
- Coding on Pico—first steps
- Soldering headers to Raspberry Pi Pico

## Technical requirements

To get going on this section, you will need the following:

- 1 Raspberry Pi Pico
- A *USB Micro cable* (either *A to Micro* or *C to Micro* depending on your computer)
- A computer/laptop running Windows/Linux or macOS
- The **Mu** editor—we'll show you how to get this
- A soldering iron and heatproof stand
- A well-lit, well-ventilated space
- Goggles

- Solder

- Solder-wick

- A breadboard

- Some male breakaway pin headers

All the code examples for this example can be found on GitHub at `https://github.com/PacktPublishing/Robotics-at-Home-with-Raspberry-Pi-Pico/tree/main/ch-02`.

# Getting CircuitPython onto Raspberry Pi Pico

For us to get going and write code on Raspberry Pi Pico, we need to put the CircuitPython interpreter on it or "*flash it*" with CircuitPython.

First, you can find CircuitPython downloads at `https://circuitpython.org/downloads`. This page shows just how many different boards support CircuitPython—although they won't support some of the unique hardware capabilities of Pico (such as PIO), it means that many of the skills learned in this book along with the code you write can be directly translated to a huge number of other boards!

Click on **Pico (By Raspberry Pi)**, and on the right is the current **stable** download of CircuitPython. There are many languages selectable here, and you can select *CircuitPython with error* messages in your language. Hit the **Download .uf2 now** button. Take note of the version number you downloaded.

Plug one end of the USB cable into your laptop. Looking at Pico, there's a tiny button on it:

Figure 2.1 – Raspberry Pi Pico BOOTSEL button

This button, shown in *Figure 2.1*, is labeled **BOOTSEL**. Hold this down as you plug the USB cable into the computer. This puts Pico into a mode for flashing firmware onto it.

You should see a new RPI-RP2 drive appear on your computer. This is Raspberry Pi Pico. Copy the Adafruit CircuitPython uf2 file from your downloads to that folder. A simple drag and drop file copy will work here.

Raspberry Pi Pico will reboot, and the drive will disappear momentarily. It will then come back as CIRCUITPY.

You have downloaded and flashed CircuitPython on Pico. This Raspberry Pi Pico is running CircuitPython and is ready to program. Next, we'll download the libraries for it.

## Preparing the CircuitPython library for Pico

CircuitPython is a good starting point—it gives you the basics that we will be using—but we will also be interfacing with other hardware. The **CircuitPython library**, united across many devices with the same version of Pico, creates an interface you can take with you to other microcontrollers should you want to try others out.

Let's use the following steps to prepare a module from the library:

1.  Open the CIRCUITPY drive on your computer and find a folder called lib. This is the target for libraries.

2.  Download the CircuitPython Library Bundle from https://circuitpython.org/ libraries. The version you download should match the version of CircuitPython you downloaded before.

3.  This gets you a ZIP file. Expand the ZIP contents, and you should get a folder with a README, examples, and a lib folder. We will keep this handy. When we need libraries from here, we copy them over to Pico.

4.  The whole library is too large to fit at once on a single Pico, but since you couldn't use all those hardware peripherals at once, you would only copy what is needed.

5.  Open the lib folder in the Adafruit Library folder, and you should see adafruit_ vl5310x.mpy. Copy (drag and drop) this file into the CIRCUITPY/lib folder. This is all it takes to install libraries for this device—some devices require a group of files to be copied over.

You've now got a copy of the CircuitPython library handy and seen how to install a module from it on Pico. We will be using this library later. We have CircuitPython and a library. Next, it's time to try some code on Pico.

## Coding on Pico – first steps

Writing and testing code on Pico is made easier with a handy tool, **Mu**. We will get you up and running and get stuff going in Pico **REPL**. We'll then write some code in a file and upload that program so that it runs when Raspberry Pi Pico boots. How do we get Mu? Let's find out in the next section.

## Downloading the Mu editor

The Mu editor gives easy access to the CircuitPython REPL. It also has a Python editor, allowing you to see the code and results together. It's small and supports other hardware-oriented Python platforms.

To download it, do the following:

1.  Go to `https://codewith.mu/`. Use the **Download** button to get the right version for your computer and install it.

2.  Launch Mu editor, and when it is running, click on the **Mode** button. From this, select **CircuitPython**. Look at the following screenshot:

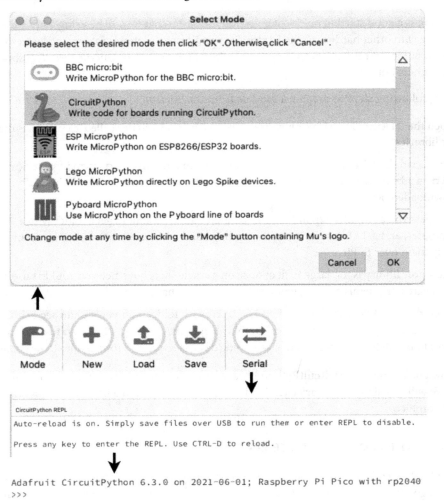

*Figure 2.2 – Mu editor buttons*

The preceding screenshot shows the toolbar in the middle, with the result of clicking the **Mode** button above it, and the result of clicking the **Serial** button below it.

3. You'll need **Serial** enabled to see and interact with Python on Pico, so click the **Serial** button as indicated. **Serial** lets you send and receive text from Raspberry Pi Pico via the USB cable.

4. To start interacting with Pico directly, click in the **Serial** window, shown below the **Serial** button in *Figure 2.2*. When you type or press keys here, they are sent directly to Pico. Pico is waiting for you to interact with it.

5. On your keyboard, press any key to start the REPL. You will see a Adafruit CircuitPython message, as shown at the bottom of *Figure 2.2*. We can start typing code here.

You have loaded the Mu editor and connected it to CircuitPython on Raspberry Pi Pico. By loading the **Serial** monitor and pressing any key here, you've entered the REPL where you can type code and see it evaluated immediately. Next, we explore some things we can do in this REPL.

## Lighting the Pico LED with CircuitPython

The console here is running CircuitPython, and it is on Raspberry Pi Pico. You can type code here. The simplest thing we can do is a `"Hello, world!"`. Type the bold text, and it should respond as follows:

```
>>> print("Hello, world!")
Hello, World!
>>>
```

Well done—this is the first bit of working CircuitPython code. However, this is about hardware, so we will light an LED instead. Type the following code into the REPL:

```
>>> import board
>>> import digitalio
>>> led = digitalio.DigitalInOut(board.LED)
>>> led.direction = digitalio.Direction.OUTPUT
>>> led.value = True
```

To talk to hardware, we start by importing some libraries.

The `board` library stores board interface details—naming and mapping pins on the device to names. What is cool about this library is that if you pick up a different board and put CircuitPython on it, then to some degree, named pins such as `board.LED` will work on that too.

The `digitalio` library has the basics for reading and writing to IO pins, defining them as a **digital pin** (versus other ways you can use a Pico pin), and then setting their direction.

We then initialize a digital pin using `board.LED`. This uses the built-in LED. That LED is on *GPIO pin 25*. You could also have used `board.GP25` for the same result. We set this pin's direction to `OUTPUT`; otherwise, you are not allowed to set its value.

Finally, in this code, we set `led.value` to `True`, which turns the LED on:

Figure 2.3 – The LED lit on Raspberry Pi Pico

You should see the LED light at this point. We should turn it off when we are done:

```
>>> led.value = False
```

You've got an LED to light—some hardware interaction. You can flip that value from `False` to `True` and back again to turn the LED on and off.

You could set any pin to `True` or `False` this way to control simple hardware, and you've used the REPL to try stuff out. However, we don't really want to do it manually— so, can we make it more automatic?

## Blinking the LED with code

The first automatic behavior would be to make the LED blink. In hardware circles, this is code is known as *blinky* and is equivalent to `"Hello, World"` for microcontrollers. This time, instead of typing this code in the REPL, we are going to write it in the editor and upload it to Pico. That way, we can tweak it and send the code again.

Start by clicking in the code area:

```
untitled ● ✖
  1   # Write your code here :-)
  2
  3
```

Figure 2.4 – Click here to start coding

As the box says, we can write out code there. To blink an LED, the basic idea is to turn it on, wait a bit, turn it off, wait a bit, and repeat. Let us see that in code:

```
import time
import board
import digitalio

led = digitalio.DigitalInOut(board.LED)
led.direction = digitalio.Direction.OUTPUT

while True:
    led.value = True
    time.sleep(0.5)
    led.value = False
    time.sleep(0.5)
```

This code starts with the import of time, which lets us control time! Well, OK—it lets us wait for a bit with sleep.

We import board and digitalio and set up the pin and its direction as before.

To make the LED keep on blinking, we put it in a while True loop, which will repeat the code indented under it until we (hard or soft) reset Pico.

Inside the while loop, first we turn the LED on, then we sleep for half a second. The time here is written as a decimal number—it is always in seconds, and we want less than a second.

We turn the LED off again, sleep again, and the code will loop around.

Save this code to Raspberry Pi Pico (the USB drive named CIRCUITPY) as code.py. It will automatically run the file named code.py—the LED should now start blinking.

While this seems simple, changing digital IO pins and sleeping is the basics of controlling motors too. In this section, you've written code to automatically blink a light, and seen how to use time and how to upload a file to Raspberry Pi Pico instead of typing everything at the REPL. Next, we'll need to prepare Raspberry Pi Pico for plugging in other hardware by soldering on headers.

## Soldering headers to Raspberry Pi Pico

Raspberry Pi Pico can run some neat code, it can blink that LED, and you could get it to input things, and print on that serial console. But it's going to be lots more fun if we start plugging stuff into it! To do that, it will need **headers** soldered into it. **Soldering** may seem daunting the first time around, but with practice, it will be a skill you'll use repeatedly in robot building.

We are going to be soldering header pins into Raspberry Pi Pico. These are breakaway pin headers:

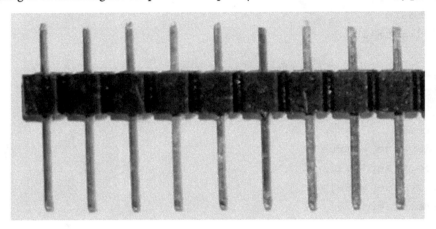

Figure 2.5 – Breakaway pin headers

Pin headers will let us plug Pico into a breadboard so that we can connect it to other electronics or use Pico with female cables to link to sensors. There are other kinds—female kinds, not so useful for the breadboard—and you can get 2-row pin headers, which you'll see in other Raspberry Pi models.

To get the right number of pins, we need to measure and snap them off:

Figure 2.6 – Snapping off the headers

Breakaway headers such as these come in strips designed to break away. One straightforward way to get the right length, instead of counting each one out, is to place the short end of the header loose into Pico, as shown in *Figure 2.6*, and then break at the join just past there. You can either cut with a knife or just pull them off at an angle. They come apart easily. We'll want two sets like this.

Raspberry Pi Pico comes with two rows of headers, and we can use this to our advantage. Push the Pico and header assembly into a breadboard:

Figure 2.7 – Raspberry Pi Pico propped onto the headers

When you solder, you do not want the headers to wobble in their sockets—you want them to stay still. The breadboard and headers as pictured in *Figure 2.7* will hold Pico up and stop it from moving.

For this work, I recommend a chisel tip for the soldering iron. Chisel tips have a good surface area to heat joints with and are common.

You should have an iron, a stand, a tip cleaner (I prefer brass wire), and some solder. I recommend safety goggles too. You should also be working in a clear, well-lit, and well-ventilated space—use a solder fume extractor if you can; you should not inhale these fumes.

It's now time to heat up a soldering iron. Allow the iron to heat up—this may take a few minutes.

> **Important note**
> When the soldering iron is hot, only hold it by the plastic grips. You must never touch the metal parts or the tip as these can cause serious burns. Always place the iron back into its stand when you put it down.

You can test if an iron is hot by touching some solder to it—when hot enough, it should melt and *wet* the iron tip. This is known as **tinning** and ensures good heat transfer between the iron and the items to be soldered. If there is a bit of a blob of solder, use the brass-wire tip cleaner to wipe it off. What you should be left with is a thin layer of solder around the iron tip. A tinned and clean tip will make far better solder joints than a dry or dirty tip!

See the next diagram for how to make a solder *joint*:

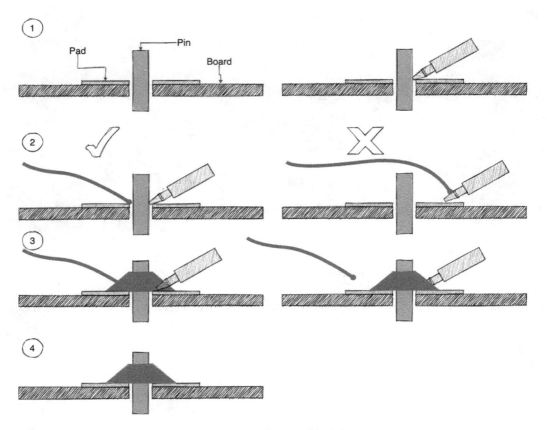

Figure 2.8 – Making a solder joint

The diagram in *Figure 2.8* shows the steps for making a good solder joint. Follow the points along with the diagram:

1.  Shown here is a pin going through a board and the solder pad (the metal bit on the board). Use the soldering iron tip to heat both the pad and the board. We want them both hot enough to accept solder.

2.  When the pad and pin are hot, push the solder into the pad and pin, not into the soldering iron. It is important for nice joints that you heat the pad and pin, then bring the solder to it—this means the solder will melt against the pad and pin, not the iron, which will lead to a good bond, electrically and mechanically. You do not need to use any pressure from the iron, and at the right temperature, the solder will flow—it does not need any pressure either.

3.  The solder starts to flow and should flow onto the pad and around the pin. You do not want a lot of solder here! Use enough to make the cone shape shown here. Remove the solder as soon as enough has flown. Pull the iron away from the joint.

4.  A good shape should be a shiny cone like this, contacting both the pad and pin. If you get a ball here, heat the solder, pad, and pin again—molten solder will only stick to hot stuff. If there is a big blob, you can use solder wick to remove excess.

We can take a closer look at how this appears in the real world on Raspberry Pi Pico:

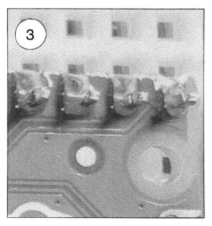
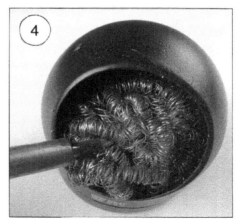

Figure 2.9 – Solder joints on Raspberry Pi Pico

The order in which you make the joint is important. The panels in *Figure 2.9* show where to start soldering:

1.  Start in one corner. In this photograph, I am heating up the pad and pin first. Ensure the pin headers and Raspberry Pi Pico do not move relative to each other when the solder cools. It will cool very quickly, but any movement here will make for a poor-quality joint. The breadboard holding strategy here should be enough to prevent most movement.

2.  Then, move across to the opposite corner. These two joints secure Pico to the breadboard, making the board less likely to move in the rig. In the photograph, I am adding solder to the joint.

3.  Once you have done two opposite corners, solder the remaining corners, and then you can start filling in the rest of the pins. Good shiny joints should look like these. There are 40 in total, but you get into a rhythm once you've done the first few.

4.  Between pins, it's a good idea to occasionally clean the tip. Use the brass tip cleaner (stab the iron into it) to clear excess solder from the soldering iron tip.

While soldering, it is possible that you will accidentally bridge two pins—that is, connect between them. This must be remedied, or you could damage your Pico when you plug it in. See the following diagram on how to clear this up:

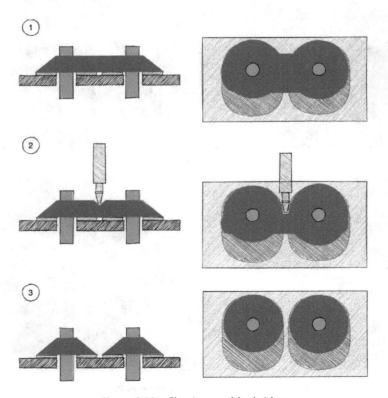

Figure 2.10 – Clearing a solder bridge

To clear a **solder bridge**, first, clean the soldering iron tip, then follow the steps in *Figure 2.10*:

1. This is the solder bridge. The solder has created an accidental connection between two pins that should not be connected.

2. Put the iron between the pins, and when the solder melts, draw the iron through between them. The solder should flow around the two pins.

3. The pins should look a little like this when the bridge has been cleared.

Now you have made the joints, inspect them to ensure that they are all conical tents, none is a bridge, and none has been missed. If so, you have completed this soldering job. Do not forget to turn off the soldering iron!

With the pin headers soldered in, this Raspberry Pi Pico is ready to connect to a robot.

## Summary

In this chapter, we've been on a close tour of Raspberry Pi Pico. We've put CircuitPython on the board and prepared some libraries to copy when we need them. We've downloaded an editor that can talk to Raspberry Pi Pico on its serial REPL, and have then written code to upload to Pico. In doing so, we've got the hardware to do some basic digital output.

We've also soldered headers onto Pico so that we can start building robots around it.

In the next chapter, we'll design the robot to build around it, using CAD and taking our test fit into a more serious gear so that we can start cutting material.

## Exercises

Try these exercises to get to know Pico more:

- Adjust the timings in the blink code. You should be able to get the light to blink quicker or slower by changing the number of seconds in the `sleep` statements.

- Could you make the light blink in an irregular pattern? You could use a series of timings. Digging a bit at CircuitPython, you could use a list of values and loop through them.

- Explore using the REPL for other Python code.

- We are going to need header pins on the motor board as well. Using the same techniques as covered in the section Soldering headers to Raspberry Pi Pico, solder pins into the motor driver board.

## Further reading

Please refer to the following resources for more information:

- At this stage, it is worth having the datasheet for Raspberry Pi Pico to hand, available at `https://datasheets.raspberrypi.org/pico/pico-datasheet.pdf`. It has a helpful pin reference on *page 5*.

- CircuitPython also has a reference guide at `https://docs.circuitpython.org/en/7.3.x/docs/index.html`. The *Core Modules* section will be helpful for further experimenting with this environment.

# 3

# Designing a Robot Chassis in FreeCAD

We've been talking about building a robot, but how do you make a **chassis**? How would you design one simple enough for a first build? Making a custom chassis takes more work than buying ready-made ones, but is a skill worth learning.

In this chapter, you will see how to use FreeCAD to make 3D designs for a chassis, its parts, and its frame. We'll consider the material we'll use and what adjustments we'll need to make to it.

You will then take this **CAD** design into the real world, making printable drawings that we'll use to cut our robot parts in later chapters. You'll be learning and using design skills for making robot designs.

In this chapter, we're going to cover the following topics:

- Introducing FreeCAD
- Making robot chassis sketches in FreeCAD
- Designing the caster placement
- Modeling chassis parts from sketches
- Making FreeCAD technical drawings

## Technical requirements

FreeCAD is free software. This chapter does not require any paid software, making it useful to a wide audience.

You will require the following:

- FreeCAD software download version 0.20 or later from `freecadweb.org`.
- Your cardboard test fit.

The FreeCAD design examples for this chapter can be found on GitHub at `https://github.com/PacktPublishing/Robotics-at-Home-with-Raspberry-Pi-Pico/tree/main/ch-03`.

# Introducing FreeCAD

FreeCAD is a free and open source **3D CAD design** tool available on most home computer platforms. In this section, we'll introduce you to the software and configure it for our needs.

If you don't yet have it, please download FreeCAD and install it before continuing. I recommend using FreeCAD in fullscreen mode if you can to accommodate all the toolbars and panels.

## The FreeCAD screen

We will start with concepts you'll need to use FreeCAD.

Let's start with an overview of the screen you'll see when you launch FreeCAD:

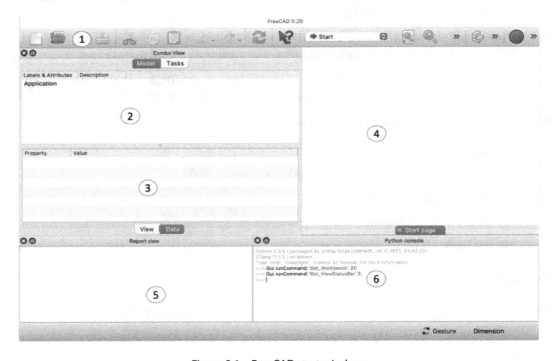

Figure 3.1 – FreeCAD start window

The preceding screenshot shows the FreeCAD start screen. The important areas are as follows:

1. The toolbar is a ribbon of buttons that changes with the workbench.

2. The **Combo** view has two modes: **Model**, which shows a model tree, and **Tasks**, which shows tasks relative to a selection.

3. The **Properties** view lets you view and edit properties. It's only visible in the **Model** view.

4. The **Main** view shows the sketch, part, or assembly you are working on.

5. The **Report** view shows output messages from the system.

6. The **Python** console shows Python code equivalent to your actions.

You'll use these views in making parts. In the next section, we will look at workbenches, which you'll need in order to use FreeCAD effectively.

## Selecting workbenches

FreeCAD has many object types—for example, parts, sketches, and drawings. A **workbench** has operations for manipulating and creating different object types. When you change the workbench, the toolbar and views change.

Let's take a closer look at **Workbench Selector** in the toolbar:

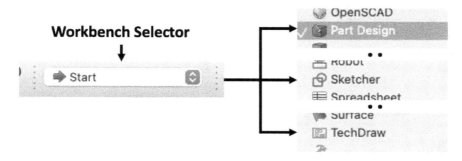

Figure 3.2 – Workbench Selector

*Figure 3.2* shows the **Workbench Selector** parts. On the left is how this selector looks in the toolbar. Clicking this will show an expanded menu of options, shown in part on the right. Highlighted here are **Part Design**, **Sketcher**, and **TechDraw**, which are workbenches we will use in this chapter. Here is a quick overview of their functions:

- **Sketcher** is where you input geometry in 2D to create parts from.

- **Part Design** is used to design 3D parts based on sketches.

- **TechDraw** makes output drawings for cutting/drilling.

> **Important note**
> Directly below the **Part Design** workbench is a **Part** workbench. They look similar. Avoid accidentally choosing **Part** instead of **Part Design**.

When using FreeCAD, be aware of the workbench you are using. The toolbar and actions you can perform in a view change based on your workbench selection.

You'll see more workbenches as you use them. We'll configure FreeCAD first.

## FreeCAD settings

We should make FreeCAD settings consistent before starting.

### Loading workbenches

FreeCAD does not start with all workbenches loaded. Therefore, we should ensure that the workbenches we will use are loaded before we continue.

You can access the **Preferences** panel via **Edit**, then **Preferences**—or **FreeCAD**—and then **Preferences** on macOS. You should see something like this:

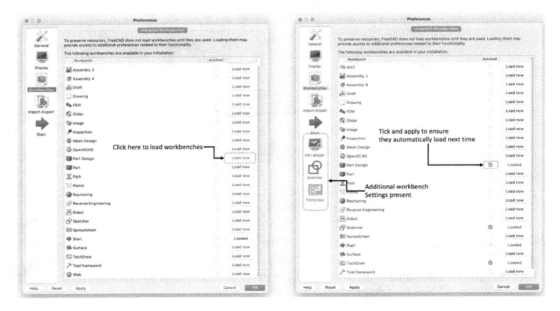

Figure 3.3 – Preferences panels: loading workbenches

*Figure 3.3* shows the **Preferences** panel, with the **Workbenches** panel selected. You can load workbenches with these steps:

1. In the **Preferences** panel, click **Workbenches**.

2. Click the **Load now** button alongside **Part Design**, **Sketcher**, and **TechDraw**.

3. Now, check the **Autoload** checkbox for each of these three sketches.

4. Press the **Apply** button. Three additional workbenches should appear.

The other preferences will make things more consistent.

### Display preferences

Display preferences ensure that your view and navigation work the same way as in this chapter. Please look at the following **Preferences** panels:

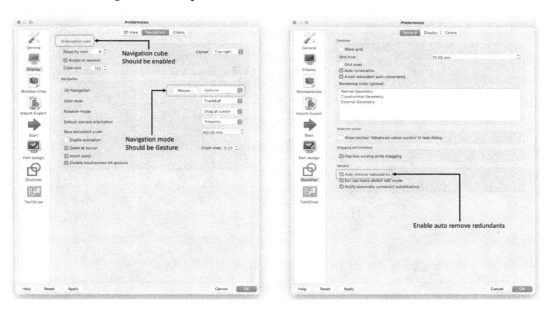

Figure 3.4 – Preferences panels

The left of *Figure 3.4* shows the **Display** preferences. Perform the following steps:

1. In the **Preferences** panel, click **Display** and then the **Navigation** tab.

2. Ensure that the navigation cube is enabled and that its **Corner** is **Top right**.

3. Set the **3D Navigation** system to **Gesture**. Gesture navigation is great for mice, trackballs, and touchpads.

4.  We also need to prepare **Sketcher** preferences. These alter how the 2D drawing mode operates. Select the **Sketcher** pane under **Preferences** to show the panel on the right of *Figure 3.4*.

5.  Ensure **Auto remove redundants** is enabled. This will remove redundant constraints and prevent conflicts later. We will explain constraints later in the chapter.

6.  Click **OK** to accept these settings.

With the settings in place, we can start creating our sketches. In the next section, we'll create a document and start sketching in it.

## Making robot chassis sketches in FreeCAD

We will model our robot, revisit the test fit as sketches, and then model it in 3D to guide us in cutting the chassis and attaching parts. We'll model parts as boxes with outside dimensions and screw holes where needed. That way, we can see where things will go without adding detail. We are aiming for this:

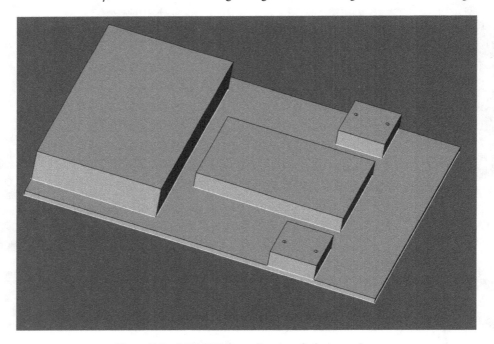

Figure 3.5 – A 3D CAD bounding box fit for our robot

*Figure 3.5* shows a model of a robot in boxes. Parts are modeled by drawing rectangles, and then pulling them into 3D. We can then use this as a guide to cut the chassis plate and then assemble our chassis.

We'll start by sketching our robot, and in later sections, we will develop it into 3D parts.

## Preparing the document

Everything you make in FreeCAD starts with making documents. The following screenshot shows how:

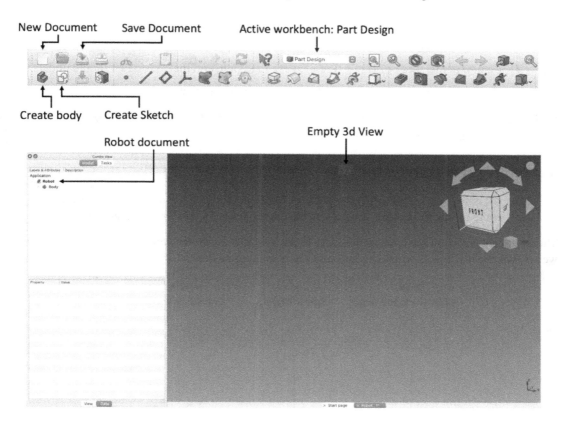

Figure 3.6 – Creating a document

*Figure 3.6* shows annotated screenshots of FreeCAD. Follow these steps to get started:

1.  Select **Part Design** from the **Workbench** menu to enter this workbench. The toolbar will look like the top of *Figure 3.6*.

2.  Create a new document by clicking the toolbar icon. You'll see an empty 3D view, shown in the lower screenshot in *Figure 3.6*.

3.  A **Part Design** 3D solid object is a **body**. Click **Create body** in the toolbar.

4.  Save this document as Robot—the name changes in the model view.

We will have a few bodies, so it's good to name them. First, ensure you have the **Model** tab selected. The default name for a body is just Body; we should rename it something more meaningful. Follow these steps to rename it:

1. Select Body by clicking it.

2. Press *Enter* to make its name editable.

3. Type a new name, SketchMain. Press *Enter* to accept it, and the body name should update.

4. Use the **Save** button or press *Ctrl/Cmd + S* to save this document.

We have an empty body, but we need to create some shapes in it. We'll do this by making main sketches for them and pulling them into 3D. Let's see how to sketch.

## Sketching the chassis outline

Sketching is a critical part of a FreeCAD workflow. We'll start with the main sketches of chassis parts and use them to model in 3D. Click the **Create Sketch** toolbar icon, as shown in *Figure 3.6*.

The document view now has a navigation cube in the top right—it shows which side of the 3D system we are looking at. The following diagram shows how to navigate here:

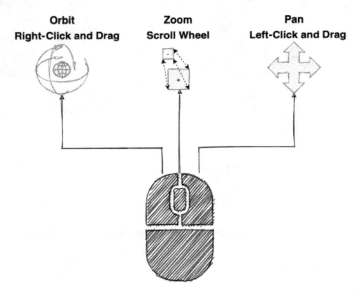

Figure 3.7 – 3D gesture navigation

*Figure 3.7* shows 3D gesture navigation. Hold the left mouse button and drag to orbit/rotate the camera around in 3D. Scroll the wheel to zoom in and out of the model. Use the right mouse button to pan, moving the camera up/down and left/right. You can click the navigation cube faces to see different views of your model.

FreeCAD is waiting for you to choose a sketch plane, as shown in the following screenshot:

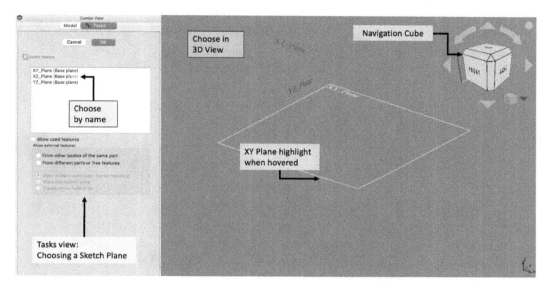

Figure 3.8 – Choosing a sketch plane

*Figure 3.8* shows a dialog for choosing a plane, a flat surface to sketch on. FreeCAD requires you to choose a surface for sketching. You have two ways to select it:

- The **Tasks** view on the left shows a list of base plane choices.
- You can choose flat surfaces or planes in the 3D view.

For this sketch, choose XY_Plane. This is the horizontal plane.

> **Important note**
>
> If you find it hard to select things, note that as you hover the mouse near FreeCAD objects, they become highlighted, and then you click to select them. This hover/click operation is vital to making selections.

Click **OK** to enter the **Sketcher** workbench, as shown in the following screenshot:

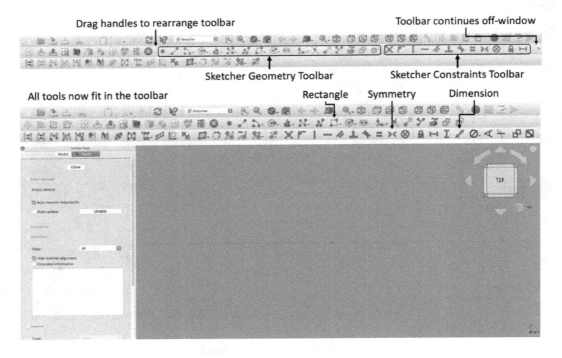

Figure 3.9 – The Sketcher workbench

*Figure 3.9* shows **Sketcher** screen elements. At the top are toolbars. We'll mostly use two toolbars—**Sketcher Geometry** to create shapes and lines, and **Sketcher Constraints** to place geometry. The default arrangement of toolbars puts items offscreen, shown in the top right. However, we can drag toolbars to rearrange them to see all the toolbars, as per the second toolbar shot.

The second toolbar shot has **Rectangle**, **Symmetry**, and **Dimension** tools highlighted. We'll be using these in the following sections.

Below this, on the left, the **Tasks** panel lists constraints and sketch components. On the right is the main sketching view. The navigation cube shows we are looking at a view of **Top**. The *x* and *y* axes divide the screen with a cross.

Let's use these tools to draw our chassis. Follow these steps:

1.  Click on the **Rectangle** tool in the geometry toolbar, highlighted in *Figure 3.9*.

2.  Create a rectangle crossing the center lines—click once for a corner and again to place the opposite corner. It should look like the following screenshot:

Figure 3.10 – Sketched rectangle

The right of *Figure 3.10* shows the rectangle. The sketch taskbar on the left has the **Constraints** and **Elements** sections. Elements are drawable items such as lines.

Sketcher uses **constraints** to position geometry. Here, there are eight constraints:

- Four have crosses with a dot in the middle; these constraints make the rectangle line ends coincident, so they join.

- The other four constrain two lines horizontally and two vertically.

Let's add more constraints to position and size this geometry.

In *Chapter 1, Planning a Robot with Raspberry Pi Pico*, we suggested 150 mm x 200 mm in our test fit. Look at the next screenshots to see how to size the rectangle:

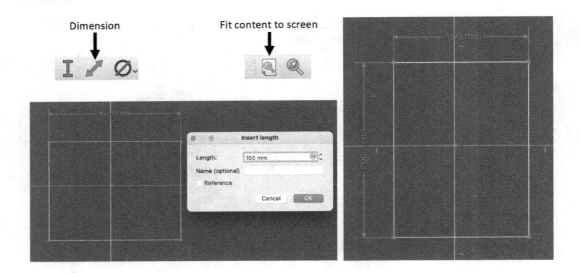

Figure 3.11 – Adding dimensions

*Figure 3.11* shows the addition of dimension constraints to the rectangle. Follow these steps to add constraints:

1.  Right-click to drop the **Rectangle** tool. Right-clicking drops your current tool.

2.  Hover the mouse over the top line so that it is yellow, and then click to turn it green.

3.  Use the **Dimension** tool in the toolbar, type 150 mm, and press *Enter*.

4.  If things don't fit on the screen, use the **Fit content to screen** button.

5.  Dimension the right line to 200 mm.

The small line icons indicate vertical or horizontal constraints. FreeCAD creates them automatically when you add rectangles.

The rectangle has dimensions but is not fully constrained, and it can move around, so we need to anchor it.

The next sequence shows how to fix it to the origin:

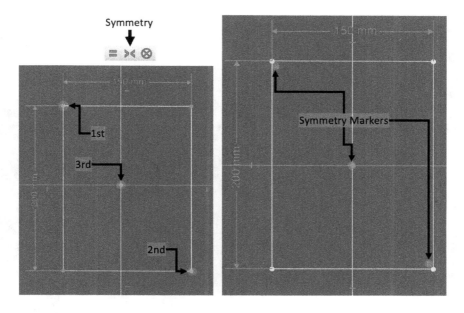

Figure 3.12 – Adding symmetry

The sequence in *Figure 3.12* shows how to add symmetry to your rectangle. The previous screenshot shows the **Symmetry** button—the left side shows selecting three points, while the right side shows a fully constrained sketch with symmetry. Follow these steps:

1.  Select the top-left point, bottom-right point, and middle point where the axes cross.
2.  Click **Symmetry** to add a symmetry constraint.
3.  When you do this, the sketch changes color to show the sketch is fully constrained. The points are locked relative to the origin.

The **Tasks** panel has also changed to show the sketch state. Look at this screenshot:

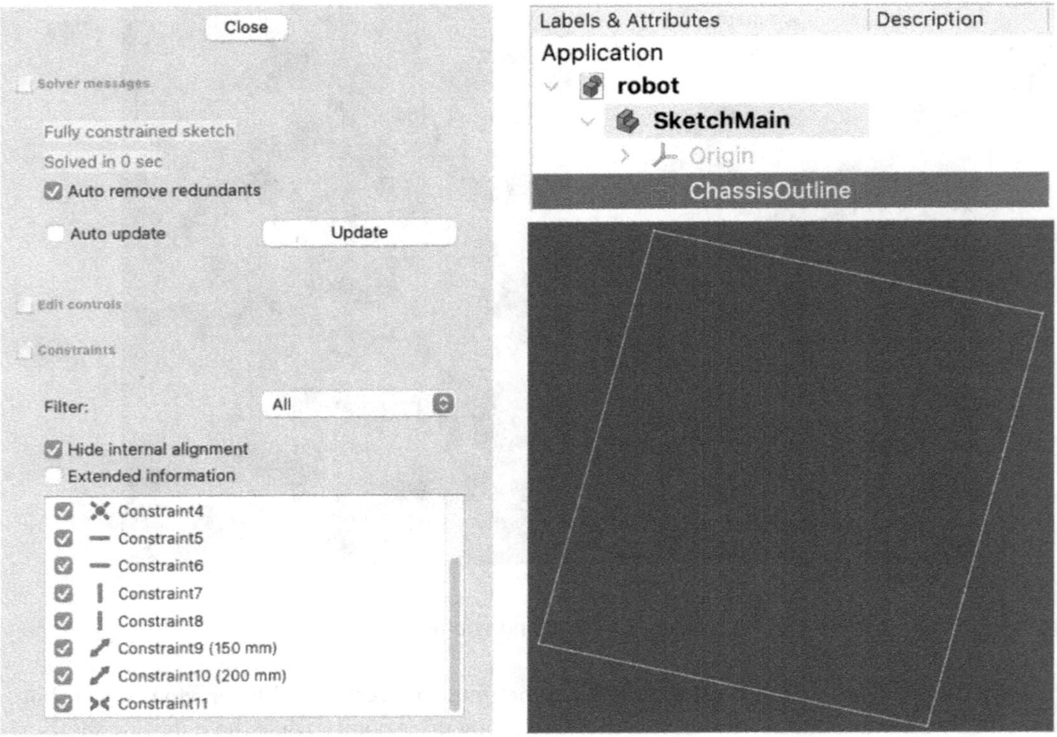

Figure 3.13 – Tasks panel for a fully constrained sketch

The **Tasks** panel in *Figure 3.13* has a **Fully constrained sketch** message in the **Solver messages** section. We have 11 constraints, including 2 dimensions and symmetry. Click **Close** to finish sketching. Name this sketch ChassisOutline.

This sketch is dimensioned and constrained. However, we will create a further sketch as the main sketch for the upper deck parts.

## Creating the upper parts main sketch

We will use a further sketch named UpperParts to make parts on the top of our chassis. It will follow the test fit but with screw holes added. It will need to reference geometry in the ChassisOutline sketch as they are related.

Create a new sketch on the same *XY* plane. We will base this sketch on the previous one by using external geometry. Look at the following screenshot:

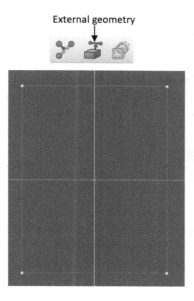

Figure 3.14 – Importing external geometry

*Figure 3.14* shows the **External geometry** tool, which you need to click, and then select all the rectangle sides to see red lines created over them. This will link to geometry from another sketch or object.

Next, we will create sketch elements for the battery box and breadboard:

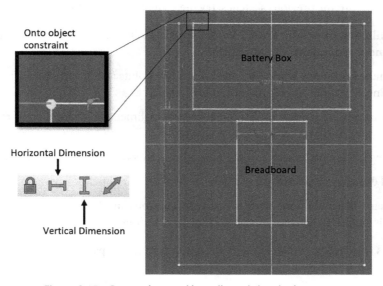

Figure 3.15 – Battery box and breadboard sketch elements

The middle screenshot shows the chassis with battery box and breadboard sketch elements on it, with a close-up screenshot on the left showing the **Onto object constraint** icon. The top-right space shows dimension tools. Proceed as follows:

1. Draw the battery box outline. By starting a rectangle top left touching the chassis rear line, FreeCAD adds an automatic **Onto object** constraint between that top-left point and the line. This makes the point stay in contact with the line. Use the test fit to set the dimensions.

2. Make the rectangle's bottom points symmetrical around the vertical axis line. The sketch should go bright green to show that it's fully constrained.

3. Next, create the breadboard. This is another rectangle, with the top two points symmetrical on the vertical axis. Use the dimensions from the test fit.

4. We want a gap between the breadboard and the batteries. Select the battery box at its bottom-left point and the breadboard at its top-left point. Make a 10 mm vertical dimension between them. The sketch should again be fully constrained.

> **Important note**
> When adding symmetry, FreeCAD should remove horizontal constraints for lines. This removes **redundant constraints**, which can conflict if constraints contradict others. Ensure **Auto remove redundants** is enabled to prevent this!

### Troubleshooting sketching

If you have trouble with the previous sketches, try these steps to solve some common issues:

- **I see conflicting constraint errors**: Ensure you have followed the preceding preference steps. **Auto remove redundants** must be enabled.

- **I can't make symmetry work**: You must select two points (dots), and then a final center dot or mid-line. Selection order is important for symmetry.

- **My dimensions are cluttered**: You can click on any dimension and drag it to move it out of the way.

- **Everything is too close to see**: Use your mouse wheel to zoom in, and then zoom out when done.

- **I am still drawing a line/circle**: If you have a tool selected, right-click anywhere to drop the tool.

- **I have items selected I don't want for constraints**: You can click on an item a second time to deselect it or click in an empty space to clear the selection.

With these parts sketched, we move on to more complicated motor parts.

## Sketching the motors

The other elements on top of the chassis plate are the motors. The motors are in our test fit as boxes, but in truth, they are a complex assembly. We are using N20 motors with built-in encoders. They need brackets to attach them.

N20 motor brackets come in two kinds—with mount holes closer to the motor or with them further apart. Since the former type is more common, we'll use these. The next diagram shows a dimensioned drawing of these brackets with motors:

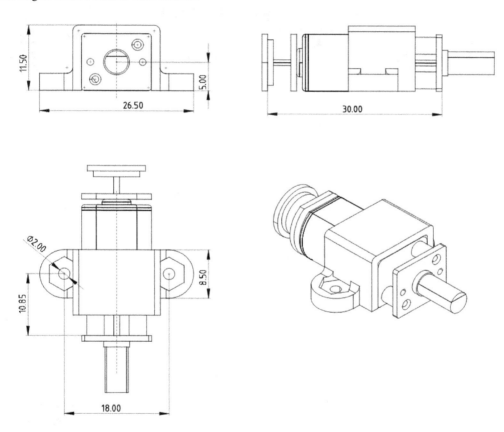

Figure 3.16 – FreeCAD drawing of N20 motor with encoder and bracket

*Figure 3.16* is a dimensioned N20 motor sketch with an encoder and bracket. We'll align the bracket front with the gearbox middle divider so that the motor won't slide.

Using this drawing, we can create rectangle outlines for the two motors, as shown in the following screenshot:

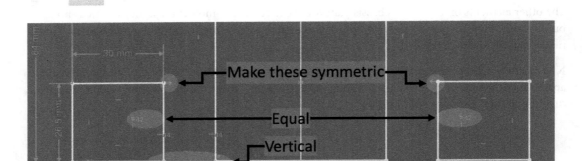

Equal constraint

Figure 3.17 – Motor outlines

*Figure 3.17* shows the symbol for **Equal constraint**, which sets line lengths or circle diameters equal. The screenshot shows the motor outline rectangles in place.

Follow these steps:

1.  Draw a rectangle on the left, ensuring that we constrain the top-left point onto the left side of the chassis. Make a similar rectangle on the right.

2.  Make the top inner points of the motors symmetrical about the vertical axis line.

3.  Select inner vertical lines on both motors, and then use the equal constraint. A little = sign shows next to the items, along with reference numbers. Items with the same number are equal.

4.  Dimension the left-hand motor using *Figure 3.16* for reference.

5.  Select the bottom-right point of the motor and the bottom left of the breadboard. Create a horizontal constraint between them to line them up.

All our rectangles are green, which shows this is a fully constrained sketch. We have sketched out the robot's top, following our test fit.

We are using more of the sketching toolbox. Now is a good time to get familiar with the tools on offer and their shortcuts. These shortcuts change, so the best way to get to know them is to hover over the buttons to get a tooltip. **Sketcher** becomes much faster when you start using its keyboard shortcuts.

A constraint reference will help; see the next table:

| Symbol | Name | Description |
|---|---|---|
| — | Horizontal | Constrains a line to be horizontal |
| I | Vertical | Constrains a line to be vertical |
| ✕ | Coincident | Constrains two points to have the same coordinates |
| ⌐ | Onto Object | Constrains a point onto an object – such as a line or curve |
| = | Equal | Constrains objects to be equal in length, radius, or diameter |
| ⋈ | Symmetrical | Constrains two points to be symmetrical about a line or another point |
| ⟋ | Dimension | Constrains the distance between two points |
| ∅ | Diameter | Constrains the diameter of a circle or arc |
| ⊘ | Radius | Constrains the radius of a circle or arc |
| ⊢⊣ | Horizontal Dimension | Constrains the horizontal distance between points |
| I | Vertical Dimension | Constrains the vertical distance between points |
| ⤙ | Tangent | Two lines, arcs, or an arc and a line must touch somewhere |
| ⫽ | Parallel | Constrains two lines to be parallel to each other |

Table 3.2 – Sketcher constraints in FreeCAD

The table shows references between symbols, names, and descriptions for each constraint type. You will see these symbols in the toolbar, main view, and constraints list on the left of the screen.

These will help create motor holes—our next section.

## Sketching the motor holes

We represent holes in sketches with circles; however, the real trick is using constraints to place them.

We can start with the motor attachment holes. The breadboard is self-adhesive and does not use bolts, and we will use hook-and-loop dots to make batteries easy to change.

Each motor has two holes. We'll make them symmetrical between the sides. Refer to the following screenshot:

## Constraining diameter

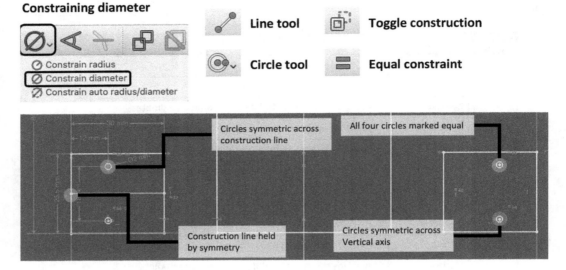

Figure 3.18 – Motor holes sketched

*Figure 3.18* is an annotated sketch of motor holes, along with the tools used. The next steps show how:

1.  First, add two circles on each side in the motor rectangles.

2.  Next, add a construction line on the left going across the motor block. Select the line and use **Toggle construction** so that it turns blue—constructions hold the geometry in place but remain internal to a sketch. Make this vertically symmetrical with respect to the motor.

3.  Make left-circle middles symmetrical across this line, and dimension them to 18 mm apart.

4.  Use a horizontal dimension to put them 12 mm in from the outside edge.

5.  Use the **Constrain diameter** tool to add a 2 mm diameter and an equal constraint for all 4 holes.

6.  Finally, make the middle of each circle on the right symmetrical to its left-hand counterpart using the vertical axis.

The full sketch should look like this:

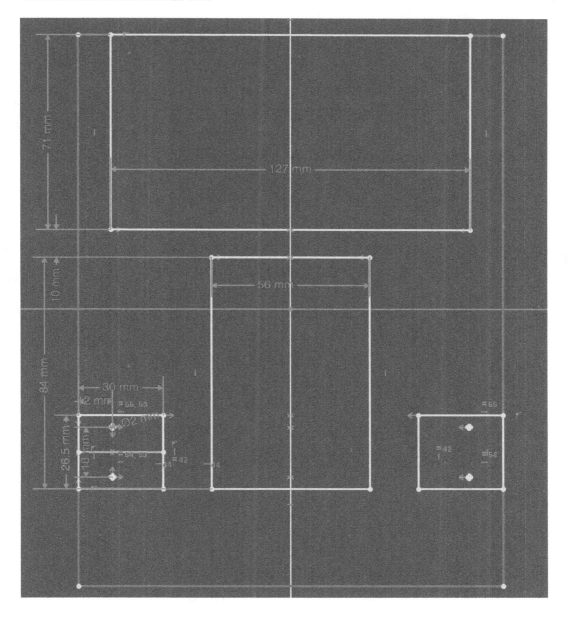

Figure 3.19 – The full chassis top sketch

*Figure 3.19* shows the whole sketch, with all components drawn on top of the chassis and their holes drawn and placed. Close this sketch.

The upper parts sketch should be 3 mm above the `ChassisOutline` sketch to account for its thickness. The following screenshot shows how:

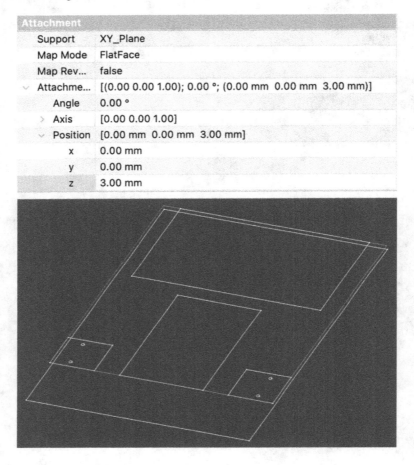

Figure 3.20 – Positioning upper parts sketch

*Figure 3.20* shows how the sketch should be positioned—first, the **Data** properties view, and then a 3D view with the sketch hovering above the chassis.

To make this, select the `UpperParts` sketch and do the following:

1.  In the **Data** tab properties tree, expand **Placement** and then **Position**.
2.  Set z here to 3 to place the sketch.

We now have modeled all the parts and their holes for the chassis top. There is also a caster for the chassis bottom. In the next section, we model this.

# Designing the caster placement

The motors and wheels are near the robot's front. The caster balances the back, making it a three-point design with two wheels to drive. Batteries are heavy, so putting the caster under and behind them will help balance the robot. We can use the battery box to guide the placement of the caster.

We need to create a LowerParts sketch to place the caster and bolt holes. I'm using a Pololu ¾ inch caster and will use its datasheet/specs as a guide. If you have a different caster, please check its specifications. Pimoroni has a dimensional drawing of this part.

The following screenshot shows the sketch we will make:

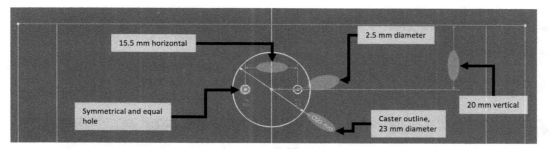

Figure 3.21 – The caster sketch

The annotated sketch shows three new circles for the caster's holes and outline. To sketch this, follow these steps:

1.  Add the rear line of the chassis as external geometry.
2.  Draw a circle for the caster outline, with its middle on the vertical axis.
3.  Using my caster datasheet, I set the caster outline diameter to 23 mm.
4.  Add a 20 mm vertical dimension between the circle's middle and rear lines.
5.  For the holes, draw a circle along this line and another circle opposite.
6.  Make the holes equal and add symmetry across the caster center point.
7.  Add a horizontal constraint between the holes.
8.  The caster information says its holes are 15.5 mm apart, and being a #2 thread, we can use 2.5 mm for the diameter. Use these to create a dimension between the hole middles and to create a diameter dimension.

We now have the upper and lower deck parts sketched. With that done and sketching skills exercised, it's time to turn these into 3D objects.

# Modeling chassis parts from sketches

We are going to start making 3D parts. The first part is the chassis plate itself, which requires two steps, and then we build other parts around it.

## Modeling the chassis plate

The following screenshot shows how we will make a 3D chassis plate:

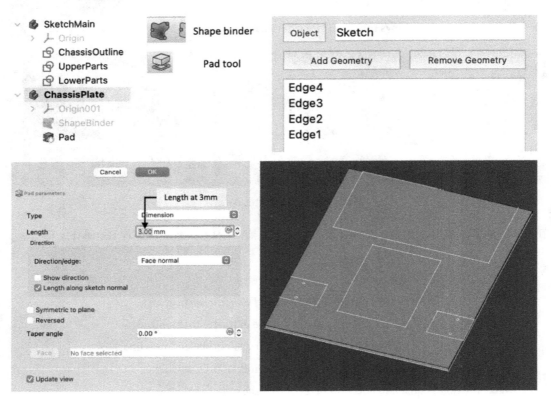

Figure 3.22 – Making the 3D chassis plate

The top part shows the **Pad** tool, the bottom left shows the **Pad Tasks** panel, and the right part shows the 3D view with the plate in it. We've put our sketches in a main sketch, so we'll need to import these. For importing geometry from sketches, FreeCAD has **shape binder** objects. Ensure you're in the **Part Design** workbench, and then follow these steps to make it:

1.  Create a `ChassisPlate` body and double-click to make it active.
2.  Click **Shape binder**, and then the task view shows the names of selected geometry.

3.  Select the four outer lines of the `ChassisOutline` sketch. Unlike sketching, you need to hold *Ctrl* or *Cmd* to do this.

4.  Then, click **Shape binder**, and then the task view shows the names of selected geometry. Click **OK** to accept this.

5.  Select the `ShapeBinder` object it created, and click on the **Pad** tool. You'll see a **Tasks** panel with parameters on the left, and a preview on the right.

6.  In the **Tasks** panel, set the length to 3 mm to match the styrene sheet. This will update the preview.

7.  Now, press **OK** to finish making the part.

This job is only partial, as we need holes too. Follow these instructions to create the upper part holes:

1.  It can help to select to hide the **Pad** we just made, using space to toggle.

2.  Select the four motor holes—hover over and highlight the circles, then click to select them.

> **Important note**
>
> In workbenches other than **Sketch**, to select multiple items, click the first item, then hold *Ctrl* (*Cmd* on Macs) and click to add additional items. If you select an item you don't want, while keeping *Cmd*/*Ctrl* held, click that item again to deselect it.

The following screenshot shows how this shape selection should look:

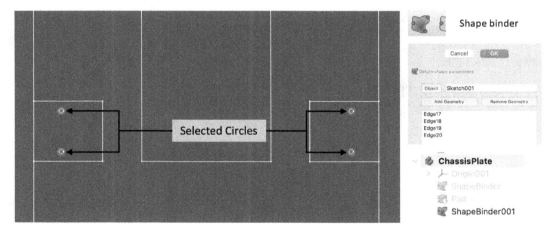

Figure 3.23 – Making a shape binder

The left of *Figure 3.23* shows holes from our sketch, all selected. The right shows the toolbar **Shape binder** icon, the **Tasks** panel for a shape binder, and the **Model** tree with a shape binder created.

You can now use this shape binder to make holes in the chassis. See the following screenshot:

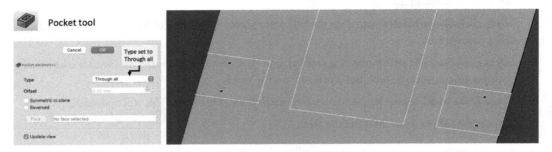

Figure 3.24 – Holes cut into the chassis

*Figure 3.24* shows how we cut holes using the **Pocket** tool in the top left. Below is the **Tasks** panel, with the cutting options. The right-hand screenshot shows the 3D output. Follow these steps:

1.  Ensure you have the shape binder selected and click the **Pocket** tool.

2.  In the dialog, select **Through all** and then click **OK**.

We've made holes to attach upper deck parts. Repeat this process for the lower part holes, creating a shape binder around the two caster holes and pocketing them through all. Next, we model the parts in 3D.

## Modeling the other parts

We make the other parts as we made the holes, using shape binders and padding them into a 3D body in the document.

The following screenshot shows how to make the battery box:

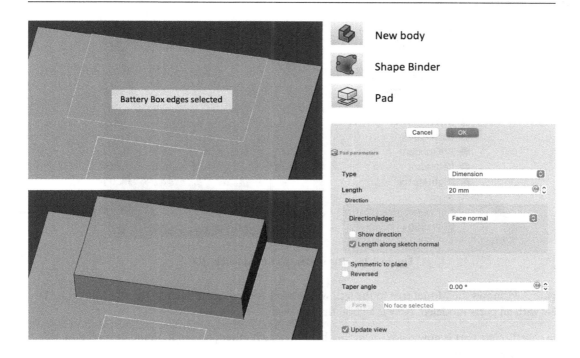

Figure 3.25 – Making the battery box in 3D

*Figure 3.25* shows how we make the battery box, with edges selected in the top left, while the top right shows the tools used, the bottom left shows the battery box in 3D, and the bottom right shows the **Tasks** panel for the **Pad** operation. Follow these steps:

1. Create a body named `BatteryBox`.
2. Select the outside lines of this box, as shown in the top-left panel. You may need to hide the `ChassisPlate` body to select all the edges.
3. Click **Shape Binder** to create a shape binder in the new body.
4. Select that shape binder, click the **Pad** tool, and set it to `20 mm`, reversed.

You should now see the preceding 3D shape. We can make other parts—motors and a breadboard—in the same way. Please see the following screenshot for the whole upper deck outline:

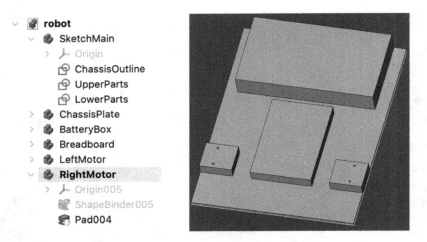

Figure 3.26 – The upper deck modeled in 3D

*Figure 3.26* shows the **Model** tree to the left, with bodies for our components. The right shows a 3D view with all parts modeled. Now is a good time to save!

The chassis plate part is now ready, including holes to attach other parts to it. In the next section, let's troubleshoot any issues you may have had.

## Troubleshooting the model

This list should help get you past common FreeCAD pitfalls with the exercises so far:

- **I can't find the right buttons in the toolbar**: Remember workbenches—when you have **Sketch** enabled, you will have different toolbars than in **Part Design**. So, ensure you have **Part Design** selected! Also, watch for offscreen buttons.

- **I clicked and missed the automatic constraint**: You can add constraints by selecting relevant items and the correct constraint tool.

- **I can't select geometry**: Use the **space** key or **Hide Item** menu to hide/unhide geometry until the item you want to select is clear.

- **Some holes won't cut**: Ensure holes aren't construction geometry. Construction geometry, colored blue in the sketch, isn't used in **Cut/Pad** operations.

- **Things aren't quite lining up**: If items are white, they aren't constrained. Use dimensions and other constraints to tell FreeCAD how they should align.

> **Important note**
>
> Sketches are best when constrained. Dragging things to *look like* they line up can lead to problems. Try to use constraints. If they are parallel, use the parallel constraint. If they are equal, use the equal constraint. If they are 10 mm apart, use a dimension.

Hopefully, this covers the issues you've had, and you have a 3D chassis with parts and holes to bolt parts to. However, we are missing the caster. In the next section, we will make it 3D using its sketch.

## Modeling the caster in 3D

The top left of the following screenshot shows pocket operation parameters for creating the caster:

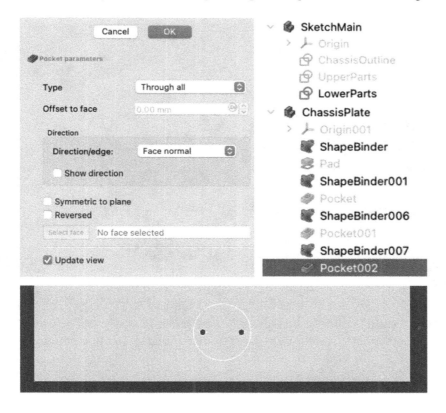

Figure 3.27 – Making the caster holes

The preceding screenshot shows elements for making the caster holes. The top left shows the **Tasks** panel – where the **Reversed** and **Through all** fields are checked for the hole. The top right shows the **Model** tree with other sketches hidden, a shape binder, and a pocket at the end. The bottom panel shows caster holes cut into the chassis.

These tips will assist in making this object:

- Look at the bottom of the model.

- Hide the last **Pocket** item in `ChassisPlate` so that you can see the sketch circles.

- Select the two circles, make a shape binder, and then unhide the chassis again.

- Select this shape binder for the holes and make a **pocket**.

- Set the pocket to **Through all**. You may need to check **Reversed** to make the cut. **Through all** only applies to geometry in the same body, and other bodies will require further pocket operations.

We can use our sketches to make a 3D caster body, using the same process as before. Look at the following screenshot to see how we achieve this:

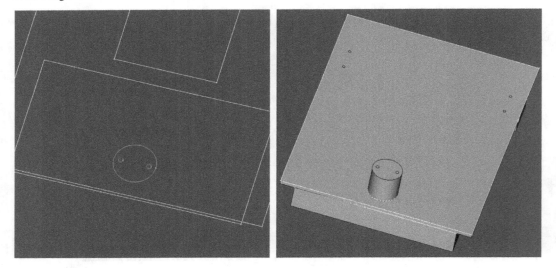

Figure 3.28 – Making a 3D caster

The preceding screenshot shows the steps in making the caster. The first panel shows the selection for the `ShapeBinder` object; the top right shows the caster padded. To add the caster, perform the following steps:

1. Hide the other solid parts. Make a `Caster` body and bring in detail from the `LowerParts` sketch into a shape binder.

2. Pad this to 23 mm, reversed. Unhide the other bodies.

We have now placed the caster and holes for its bolts. With the caster attached, we have made a rough 3D model of our robot. It's missing wheels and detail, but enough to show that this will work. We will use this design to make parts, which need a drawing—our next section will introduce these.

# Making FreeCAD technical drawings

Our design shows things will work, but we need a way to cut and drill styrene plates and rods. FreeCAD can make a **technical drawing** to help us. You've seen the *motor assembly* and *chassis base* drawings. Now, you will see how to make your own.

## Setting up the page

We'll start our drawing by setting up the page in the **TechDraw** workbench. The following screenshot shows how:

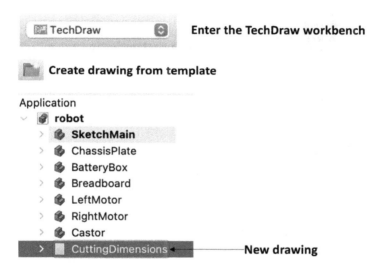

Figure 3.29 – Preparing a drawing page

*Figure 3.28* shows how we create a page for our drawing in **TechDraw**.

Choose the **TechDraw** workbench in the workbench selector, as the top of *Figure 3.28* shows. Then, proceed as follows:

1.  In the toolbar, click the **Create a drawing from template** icon.
2.  This shows drawing templates in a file selector. Choose A4_Portrait_blank and open this.
3.  This will add a new drawing to the **Model** tree. Rename this CuttingDimensions, as shown in *Figure 3.28*, left.

We now have a blank page ready for you to put your parts on. We will project parts onto this sheet—choosing parts and letting **TechDraw** draw them.

## Adding parts to the drawing

We can start by adding the chassis plate part to our drawing. Let's see how this works in the following screenshot:

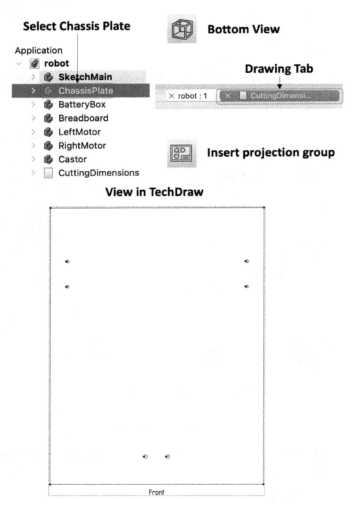

Figure 3.30 – Adding the chassis plate to the drawing

The preceding screenshot shows how we'll add the chassis plate part. The following steps describe this process:

1.  First, from the **Model** tree, select ChassisPlate as shown in the top left in *Figure 3.29*. This will take you to the 3D view.

2.  In the 3D view, use the **Bottom view** toolbar button or navigation cube to look at the bottom view of the model.

3.  Use the tabs below the view (middle right) to go back to `CuttingDimensions`. `ChassisPlate` should remain selected in the **Model** view.

4.  Click on **Insert projection group** in the toolbar (shown middle right). You should now see `ChassisPlate` appear in the drawing.

5.  Projection frames let you drag them around in the drawing. Click and drag this frame to the right so that we can put other parts on the left.

6.  Click **OK** in the **Tasks** panel to finish inserting the projection group.

You now have the chassis plate part; however, the frames are untidy, and it would be good to add hole centers and dimensions.

## Preparing the drawing for print

Some finishing touches are needed to make this drawing ready to use as a part-cutting guide. The following diagram shows how:

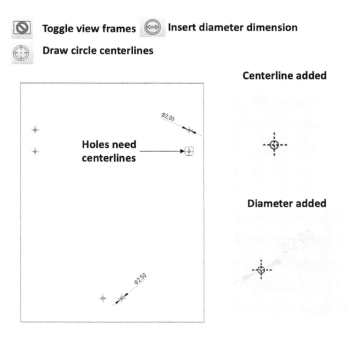

Figure 3.31 – Preparing our drawing for print

*Figure 3.30* has icons we'll use for this process along the top. These should be in the toolbar, but you may need to move toolbars to see them all. The lower left shows our part without centerlines, and the right shows the centerline added, followed by the diameter added. Perform these steps:

1.  First, we can turn off frames by clicking **Toggle View Frames** in the toolbar.

2.  So, we can drill holes, but we need centerlines. Start by selecting a circle. Then, use the **Draw circle centerlines** tool to add centerlines.

3.  We also want to add dimensions so that we know how much to drill out. Click on a circle, and then click **Insert diameter dimension**.

4.  Repeat the centerlines for all holes. Dimension the holes, but where they are symmetrical, you don't need to dimension their counterpart.

You should now have the finished drawing. You can print this off onto A4. Be sure not to scale it when printing. The printed drawing should look like this:

Figure 3.32 – The printable drawing

This printable drawing looks good. You will be able to take that printout and use it to fabricate parts, and you have learned how to project parts with **TechDraw**. We'll make those parts in the next chapter!

# Summary

You've had a brief FreeCAD tour and designed objects in it, from sketches to 3D objects with pads and pockets. You've seen shape binders to reuse geometry and used constraints in sketches to specify geometric relationships. You've used these tools to build a 3D robot design from multiple parts.

Finally, you saw how to make output drawings based on the 3D design, which you can take into a workshop with you.

In the next chapter, we will take these designs into the workshop, make parts from them, and then assemble them into a custom robot.

# Exercises

You can follow these exercises to improve and practice your FreeCAD skills and improve the robot model.

The next screenshot shows a suggestion to add a little embellishment to this robot. This exercise is highly recommended.

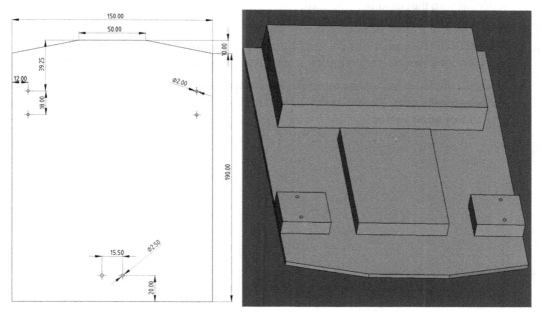

Figure 3.33 – Detail on the front of the chassis

The previous screenshot shows a drawing to the left with dimensions to cut some angled detail into the front chassis. Note that measurements have been used instead of angles, which should keep it simple to manufacture later. The right shows a 3D view of this.

The robot looks quite plain with a rectangular chassis. Adding some cuts on the chassis would add detail, and we can build upon it later for sensor positioning. Use the preceding screenshot for dimensions. Aim to cut the `ChassisPlate` body with a new sketch, as modifying the `ChassisOutline` sketch is not recommended.

Some other optional reader exercises are the following:

- Consider adding wheels to the motors—they may need an axis. **Hint**: sketch on the *YZ* plane.

- Could you use your drawing skills to reproduce an assembly drawing like the one seen in *Figure 3.16*?

## Further reading

This has only scratched the surface of what FreeCAD is capable of and how you could use it. I recommend reading further about its use at the following sources:

- *FreeCAD [How-To]*, by *Daniel Falck* and *Brad Collette*, published by *Packt Publishing*—Solid modeling with the power of Python. This tours the workbenches, multiple modeling techniques, and using Python to work in FreeCAD.

- The FreeCAD forums at `https://forum.freecadweb.org/` have a *Help on using FreeCAD* topic, which is a great place to ask questions when having FreeCAD issues.

- FreeCAD also has a wiki, with an excellent page on the **Sketcher** workbench at `https://wiki.freecadweb.org/Sketcher_Workbench`, and a *Getting started* topic at `https://wiki.freecadweb.org/Getting_started`.

# 4

# Building a Robot around Pico

Building a robot involves utilizing some practical skills in a workshop – cutting, drilling, and, most importantly, measuring. It also requires assembling parts and keeping them in place. The robot platform will be the base for our Raspberry Pi Pico robotics experiments, since we want to try out our FreeCAD designs in the real world.

Cutting and scratch building from styrene is a workshop technique that's used by model makers and robot builders in many situations – it's cheap and convenient in terms of materials but requires some patience and practice. Could you 3D-print, CNC-mill, or laser-cut these parts? Almost certainly – however, not everyone has access to these tools, and learning about scratch build techniques to complement them will give you flexibility as a workbench wizard.

Later in this chapter, we will look at wiring our robot and connecting the electronics so that the robot is ready for us to code on.

By the end of this chapter, you will have built the robot base platform with a Raspberry Pi Pico at its core, and it will be assembled and wired.

In this chapter, we're going to cover the following main topics:

- Cutting styrene chassis
- Drilling out holes
- Assembling a robot chassis
- Wiring a Raspberry Pi Pico robot

## Technical requirements

This chapter uses various tools and materials. You will need to have the right tools and be able to work safely. You'll find this equipment in the shopping lists from *Chapter 1*.

You will require the following materials:

- 3 mm styrene sheet – A4 or larger
- A soldered Raspberry Pi Pico
- The solderless breadboard
- Precut jumper wire kit
- The motor controller with headers soldered in
- A 5V 3A UBEC
- 8 x AA battery compartment, with switch
- N20 micro-metal gear motors with encoders
- Ball caster ¾ inch with 4 x M2 nuts – 2 x M2 x 6 mm
- Hook and loop/Velcro dots
- Wheels with N20 d-hole hubs
- 1n5817 or equivalent Schottky Diode
- A standoff or a mounting kit with M2 standoffs, bolts, and nuts

You will need the following tools:

- A plastic cutter
- A pair of scissors
- A metal ruler that's at least 250 mm long
- Sandpaper – 400, 600, and 1,200 grit
- A pin vise drill with 0.5 mm, 1 mm, 2 mm, and 2.5 mm **high-speed steel** (**HSS**)/twist bits
- Safety goggles
- A flat work area with good lighting, free of interruptions or being nudged
- A cutting mat
- Screwdrivers with appropriate ends for the bolts
- M2 and M3 spanners to tighten the bolts and standoffs
- A multimeter

# Cutting styrene parts

Styrene is a great material for building robots. It is easy to find, can be cut and glued, and comes in many forms. We will be using sheets and rods. In this section, we will look at cutting this material, starting with transferring our CAD measurements. First, we look at making a good cut and refining the parts so that they are smooth.

## Transferring CAD measurements to a plastic sheet

Before we cut, we will need to make markings so that we know where to cut. In the previous chapter, we made a paper template and ensured that we printed it out 1:1 on a sheet of A4 paper. You'll need that, along with some tools, as shown in the following figure:

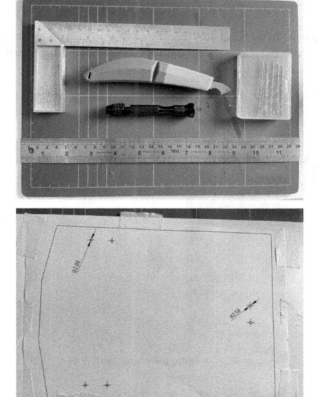

Figure 4.1 – Parts needed to transfer the drawing

The preceding figure shows the tools we will need, laid out and ready to use, followed by the drawing adhered onto a sheet of plastic. Prepare the following:

- Scissors
- Pin vise and bits
- Metal ruler
- Plastic cutter
- Cutting mat
- Sheet plastic
- Paper template
- Tape

The first thing you will need to do is cut around the drawing on the paper template. Then, you will need to tape this firmly onto the plastic sheet. Try not to let any of the edges curl – the better this is taped down, the better your result will be.

We will use this template to draw some dots on our plastic. The following figure shows this:

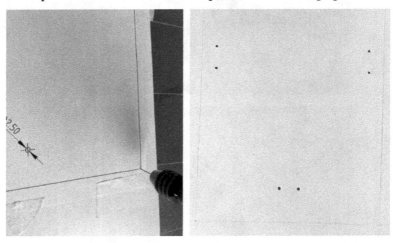

Figure 4.2 – Marking the plastic with dots

Here, you can see me using the pin vise with a bit to draw dots for each corner and hole. The right side shows the holes drilled. Follow these steps:

1. Use the pin vise with a small bit, maybe around 0.5 mm.
2. Align the pin vise over a corner on your drawing and mark through the paper and onto the plastic.
3. Repeat this for all the corners. There are six for the chassis.

4.  Now, do the same with the hole centers. There are six holes in the chassis plate.

5.  We can then use larger bits to drill out the holes – a 2 mm bit and 2.5 mm bit. Let the template guide you on the sizes.

While making the marking holes, and the bolt holes, it is important to go slowly and use little force. Accuracy is important. Do this over some waste wood or a cutting mat.

Now that we have traced this out, we are ready to start cutting the parts out of the sheet.

## Cutting the plastic sheet

Cutting takes patience. Go slowly and start with accuracy as the goal; there's no eraser for cuts!

> **Important Note**
>
> Your fingers or limbs must never be in the path of the blade! Do not cut yourself. When you are holding the ruler, keep your fingertips in the middle of it.
>
> Ensure that the plastic cutter is sharp and has a fresh blade. Blunt blades do not cut effectively, and using more pressure is dangerous and risks damaging the parts.

Let us start cutting. Look at the following figure:

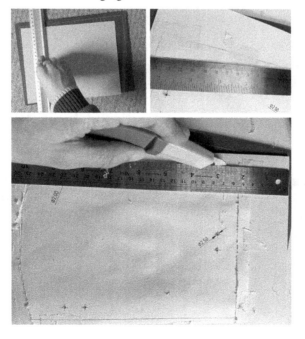

Figure 4.3 – Cutting the parts

The preceding photo shows me making cuts in the plastic sheet. The top left has me aligning a metal edge for the rear cut, and the top right has me setting up a cut for the front bevel. The bottom drawing shows the cut being made. I have a process for cutting these:

1. Line up the ruler with a pair of holes and the template so that the cutting edge faces *out* from the material you will keep. This means that any runout from the knife will not damage the part. Keep the ruler firmly held so that it doesn't move.

2. With the cutter, you will use the point of the hook.

3. Place the hook just in from the hole (so it doesn't snag on it). With light pressure, score the sheet following the ruler to the next dot. Do not aim to cut deeply – score the sheet. Do not force this – don't damage the ruler, and focus on keeping the cut straight.

4. We aim to cut about halfway through. To make a clean cut, ensure that you take things slowly. When you reach the end, keep the ruler in place, return, and follow the existing score. As you make a few passes, you will see the hook picks up more material. This is tedious but leads to a good finish.

5. Clean the blade if it gets gummed up and change it if it's dull.

6. Be patient; if you go too fast or use too much pressure, then you may wander off and damage the part or, worse, ride up and take a chunk out of you.

7. Repeat this process for all the lines you have drawn.

8. Rotate the part and ruler to get a good cutting edge for yourself. Do not lean over in awkward poses.

At this point, you should have scored all the lines for the parts, but they will still be mostly inside the plastic sheet. So, let us see how we get them out:

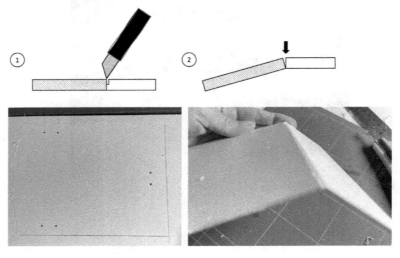

Figure 4.4 – Pushing out the parts

The preceding figure combines two diagrams and two photos. The process for pushing out the parts depends on having made enough passes that you have a deep knife score, as item **1** shows. The more knife passes you made previously, the easier this will be.

Item **2** shows what is going on; we flex the plastic at the scored line until it snaps. The bottom-left picture shows all the lines deeply scored in the sheet, and the bottom right shows the sheet being flexed to release this part. This process is very satisfying!

Once you have snapped out the parts, you should have the main chassis plate:

Figure 4.5 – The plate cut out

The preceding figure shows the chassis plate on a cutting mat. It looks a bit rough, though, with burrs on the holes and the edges. We will fix this in the next section.

## Finishing and sanding the chassis plate

The first things we will clean up are the holes. Burrs are the spiky ridged bits that remain from drilling. They can be sharp and catch your fingers and may also prevent parts from being assembled well. Let us learn how to remove them:

Figure 4.6 – Cleaning up the holes

The first panel shows two holes – one with the burrs visible and the other with the burrs removed. To do this, use a craft knife to make a circle with the blade lightly inside the hole; this will just take off the burr without widening it. Again, use no pressure here; let the blade do the work.

Repeat this for all the holes. If you need to clean the blade, use tissue, but never put your fingers in the path of the blade!

### Sanding the parts

The following figure shows how to sand this part so that it is ready to use:

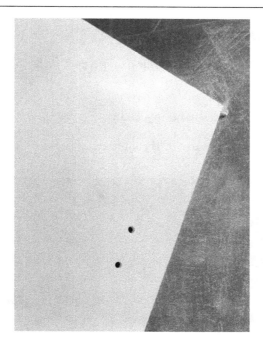

Figure 4.7 – Sanding the plate

The preceding figure shows how to sand the chassis. These tips will help:

1. Place the sandpaper on the table and stroke the part against it.
2. Start with the roughest (400 or 600 grit) sandpaper.
3. Smooth the edge and take off any ridges left by the cutting. It helps to come in at an angle of 30 to 45 degrees.
4. For long edges, hold the plates in one hand and stroke the sandpaper against them. Keep this light and support the paper where it contacts the part.
5. Now, go from the low-grit sandpaper to the medium-grit sandpaper (800–1,000) and then to the high-grit sandpaper (1,200) with the lightest touch.

The part's edges should now be smooth and not sharp to the touch.

With that, you have designed, cut, and sanded your chassis base. Sure, it could be more exciting in terms of its shape, more robust, and extend up into three dimensions or more, but you now have the skill to enhance it in these ways. A chassis base is not enough for a robot, though, so we'll start attaching robot parts to this base in the next section!

# Assembling a robot chassis

The chassis is the base of our robot. We have cut the required part, preparing it for use. In this section, we'll attach the caster battery box, and motors.

## Attaching the caster and battery box

The battery box is assembled above the caster. Therefore, we must start with the caster; the following figure shows how:

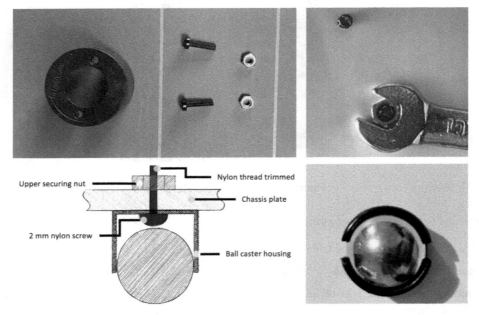

Figure 4.8 – Attaching the caster ball to the chassis

The preceding figure shows me attaching the ball caster to the chassis. Let us take a look at this in more detail:

1.  The large inset at the bottom left shows a cross-section of how this will turn out. Note that the nylon screw must go through the bottom of the ball caster housing, through the chassis plate, and then an upper securing nut.

2.  The top-left figure shows the parts for this assembly. We have the ball caster housing (with the ball separated), 2 x M2 long nylon screws, and 2 x M2 nuts. Also, prepare a suitable screwdriver, spanner, and wire/side cutters.

3.  After screwing both lower-spacing nuts into place, push the protruding threads through the corresponding holes in the underside of the chassis plate.

4.  Tighten the securing nuts with a spanner.

5.  To finish, trim the threads so that they are 1 mm or less above the securing nuts to ensure they will not foul the bottom of the battery box. A flush cutter/side cutter is a good way to do this without destroying the thread.

6.  Pop in the ball caster. Check that the ball will move freely and isn't catching on the screws.

At this point, the ball caster is secured. We can now attach the battery box, as shown here:

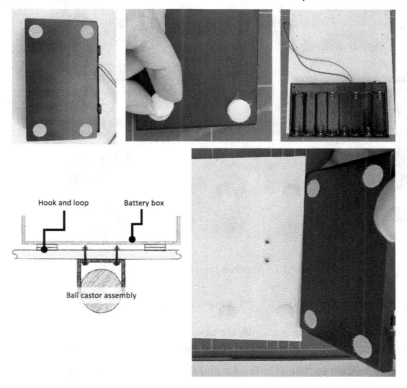

Figure 4.9 – Attaching the battery box

*Figure 4.9* shows how to attach the battery box. The top left shows hook and loop disks being adhered to the battery box in the four corners. Next, we add their opposite disks onto them. We then align the battery box over the chassis and push it down firmly so that the bottom disks stick to the chassis. Leave this for a few minutes for a good bond.

The bottom left shows a cross section with the relationship between the battery box assembly and caster underneath. The final panel, at the bottom right, shows that once the hook and loop disks are adhered, you should be able to pop the battery box off the hook and loop to make changing batteries easier.

Now, let us move on to the robot's front, which is where the motors will be added.

## Attaching the motors and wheels

Now, let's bolt the motors on and fit their wheels in place. The following figure shows the necessary steps:

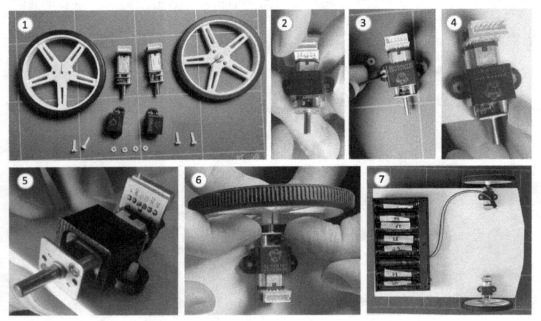

Figure 4.10 – Attaching the motors and wheels

The preceding figure shows how to attach the motors and wheels:

1.  The top-left panel shows the parts that are needed – two wheels, two motors, two N20 motor brackets, four nylon bolts, and four nuts (which should have come with the N20 motor brackets).

2.  Fit the motor bracket so that it covers the middle gearbox plate and part of the motor.

3.  Pop the nuts in the side lugs – a small jeweler's screwdriver will help here.

4.  The motor will line up over the holes. It may help to put the screws into one of the sides first and slot the motor over them.

5.  Tighten the screws into the nuts on the motor bracket. Repeat *steps 4* and *5* for the other motor.

6.  Attach the tires to the wheels – these are usually separate. Carefully push the wheels onto the motor spindles. Ensure that pressure is being placed on the motor but try not to flex the chassis.

7.  We have now assembled the robot chassis, other than the electronics.

We've assembled the chassis enough for it to stand on its wheels. It has a battery box and motors, but nothing to instruct the motors when to move. In the next section, we will wire the robot's electronics so that we'll be ready to run the necessary code.

# Wiring a Raspberry Pi Pico robot

In this section, we will look at connecting Raspberry Pi Pico to the motors via the motor controller we chose in *Chapter 1*, *Planning a Robot with Raspberry Pi Pico*. We will add the power circuitry and recommend a few techniques for robot wiring. Although the breadboard can adhere to the chassis, I tend to leave off using adhesive until it's necessary – it is easier to wire and make changes sometimes with the breadboard out of the robot.

## Wiring Pico and the motor controller into the breadboard

The motor controller that we suggested in the planning phase was the TB6612-FNG. I have used a SparkFun board. The following table shows the pins for this module:

| Pin name | Type | Function |
|----------|------|----------|
| GND | Power | Ground – power connection back to batteries. |
| VM | Power | Voltage for motors – higher than the rest of the system. |
| VCC | Power | The logic voltage should be at 3.3v, the same as Pico. |
| AIN1/2 and BIN1/2 | Input | These are the motor control input pins. Raspberry Pi Pico will control the motors through these. A and B designate sets of motor outputs. |
| PWMA/B | Input | These can be pulsed on and off to control the motors. We will tie these to the VCC voltage so that only the AIN/BIN pins are needed. |
| STBY | Input | Connect this pin high (VCC) to enable the motors. |
| AO1/2 and BO1/2 | Output | These are the motor output pins and can be connected directly to each motor. |

Table 4.1 – TB6612-FNG pin functions

The preceding table describes each pin, its overall type, and its functionality.

We should also use the datasheet for Pico pinout at `https://datasheets.raspberrypi.org/pico/Pico-R3-A4-Pinout.pdf` while considering the pins we'll use. The following diagram shows how we will wire this controller:

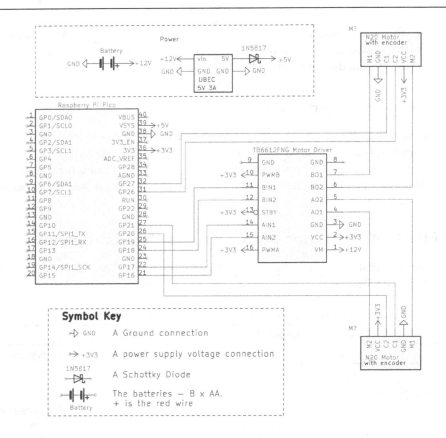

Figure 4.11 – Wiring diagram for the robot

This wiring diagram (made in KiCad) shows the wiring we'll be using in this chapter. The box at the bottom shows the symbol key.

Focusing on the connections between Raspberry Pi Pico and the motor module, you'll see the **General Purpose Input/Output (GPIO)** pins wired to control the AIN and BIN pins. The preceding diagram shows the Raspberry Pi Pico 3v3 pin connected to a +3v3 power line, which we've also connected to the motor controller's VCC, PWMA/B, and STBY pins.

We will build our circuits on a breadboard. With pre-cut jumper wires, you can push the exposed metal ends of the wire into the breadboard holes to make a connection. The top and bottom have power rails for connecting the power and the ground. The middle section has 30 rows, each with 2 connected rows consisting of 5 pins.

How do we fit this on a single breadboard? Let us look at the following diagram:

Figure 4.12 – Wiring the motor connections on a breadboard

The preceding diagram shows a suggested breadboard layout. First, we must plug in Raspberry Pi Pico and the motor module, with Pico's USB connector facing away from the motor board. Next, we must prepare the power rails, connecting the black rail to GND and the red rail to 3.3V – do this with longer wires so that they don't block the USB port. We must also connect the motor board's PWM and STBY pins to the red rail. Finally, we must make four connections between the GPIO pins of Pico and the input pins of the motor controller. Note the gap at the end of Pico – there is a spare ground pin here.

Make the connections with pre-cut jumper wires. Favor making it easy to inspect and change compared to tight and compact wiring.

We've wired the motor controller, but this board needs power – we'll look at this next.

## Adding the batteries

We are wiring three main power components to this robot – a battery box, a UBEC to convert the battery output into the 5V input that Raspberry Pi Pico needs, and a diode to stop USB and UBEC power interacting with each other.

The top of *Figure 4.11* in the previous section shows the UBEC and diode. How do we add those? Let us look at a suggested breadboard layout:

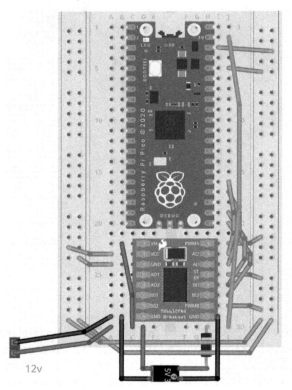

Figure 4.13 – Suggested board layout for batteries, UBEC, and the diode

This breadboard layout builds upon the previous one, with the new wiring picked out. The connection marked as 12v is for the 8 x AA battery box. The UBEC (bottom) has the *in* side coming from the battery and the motor board ground, with the 5V *out* side going through the diode into the Pico VSys pin. Also, note the wire from the battery output to VM – this is for the motor voltage on the motor board. Where the battery comes in, I suggest using a marker or label to show this since reconnecting this elsewhere can damage Pico or the UBEC.

These parts may be tricky to wire because one end of a UBEC typically has very thick wires, and the battery box wires tend to be multi-stranded and not suitable for breadboards. However, you have two options here:

- Push sets of two male header pins into a breadboard and solder wires onto them.
- For a tidier but more difficult option, you can crimp DuPont/Mini-PV male connectors onto the ends of the battery box and the input side of the UBEC. I recommend crimping these while following the guide at `https://mschoeffler.com/2017/11/02/tutorial-how-to-crimp-dupontmini-pv-connectors-engineer-pa-09-connector-pliers/`.

The output side of the UBEC is typically female. Therefore, I used pre-cut jumper wires in the breadboard to connect the 5V and ground separately.

> **Important Note**
>
> The red rails of the breadboard are running the 3.3V power output from Pico. Therefore, do not connect the 5V output of the UBEC or 12V input power to this red rail!

Our robot has power circuits, which means it can power itself. In the next section, we will wire in the motors.

## Wiring in the motors and encoders

The motors we chose use six wires for each motor. We need to be clear on what the connections from these mean. Based on the datasheet Adafruit provides for the TB6612FNG device, the following table will help you make these connections:

| Motor pin | Wire color | Purpose |
| --- | --- | --- |
| 1 – M1 | White | Motor 1 – Motor power/control |
| 2 – GND | Blue | Encoder ground |
| 3 – C1 | Green | Encoder signal 1 |
| 4 – C2 | Yellow | Encoder signal 2 |
| 5 – VCC | Black | Encoder power – 3v3 |
| 6 – M2 | Red | Motor 2 – motor power/control |

Table 4.2 – N20 motor with encoder connections

If you have purchased a different device, please use its datasheet instead.

The wire colors match the wired connectors that come with these motors. Plug those into the motors carefully, as shown here:

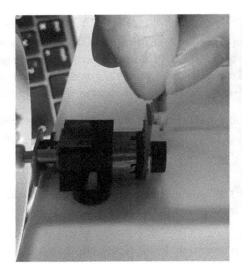

Figure 4.14 – Plugging the motor cables in

The preceding figure shows me plugging the wires into their motor sockets. These wires come with a specific set of colors. Beware that the wire colors do not correspond to conventional black/red coloring here and that I have specified the colors in *Table 4.2*.

We can now proceed to wire this. Refer to *Figure 4.11* for the overall circuit diagram. We've connected the motor power/control lines to the motor controller. We need to wire the motor controller to the motors, and the encoder signals to Raspberry Pi Pico. Note that the white and red cables – M1/M2 on the motors – are plugged into the AO/BO points on the motor controller. Next, we must connect the yellow and green encoder cables to Raspberry Pi Pico.

The motor wires are just about stiff enough to go into a breadboard as-is; however, you will have far more robust connections if you also crimp these cables. I grouped them into three pairs for each motor:

- M1/M2 motor control pins
- The green/yellow encoder pins
- The blue/black encoder power pins

Pairing them makes it much easier to plug them into their intended location on the robot.

With that, we have made all the wiring connections for the first phase of this robot. In the next section, we will look at powering it up.

# Powering the robot up

The robot has the necessary connections for this stage. The following figure shows what they look like:

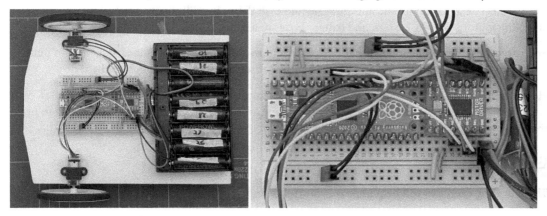

Figure 4.15 – The fully wired robot

The preceding figure shows the breadboard fully wired.

Now, check the wiring carefully, paying extra care to the power connections to look for potential short circuits in terms of any voltage to the ground or other voltage rail.

Place eight fresh AA batteries into the battery box and turn it on. You should see the light activate on Raspberry Pi Pico. If not, quickly switch it off and check the connections. If anything is warm, then this is usually a sign of a short circuit or incorrect connection. If nothing is warm, verify that the connections are not loose.

If there are no loose connections, this is where I would recommend using a multimeter to check the wires – in continuity mode. I suggest the Adafruit guide at https://learn.adafruit.com/multimeters, which explains how to use a multimeter. I would check any crimped wires, along with the voltages in and out of the UBEC.

If you find that a crimped wire is not conducting, I suggest replacing the crimp. A cheeky bit of solder in the crimp may do, but this can lead to a more fragile crimp later. You should be able to try and power the system up again.

At this point, you should be able to turn on the robot and see the LED light on Raspberry Pi Pico (and possibly the UBEC).

# Summary

With that, you have built your robot from the designs provided. You learned how to measure patterns from a CAD design onto plastic and then cut and drill them. In addition, you now have experience attaching motors, wheels, casters, and battery boxes to a robot.

You wired the robot while following the circuit diagram and suggested breadboard layouts to build. Then, you powered up the robot and performed some troubleshooting (if it did not light up).

However, the robot will not move yet, as we have not implemented any code. In the next chapter, we will add code so that we can move the robot and control its wheels. Then, we will learn how to use them to make turns and drive at different speeds.

# Exercises

The following exercises will help improve your robot and help you practice your skills:

- This robot is quite flat. Consider how you might add other layers or add depth to it.
- The breadboard is currently not adhered to the robot; perhaps hook and loop disks could help?
- It is a good idea to get familiar with the robot's wiring. Sketch a diagram for yourself showing the path between Raspberry Pi Pico and the motors.

# Further reading

To learn more about the topics that were covered in this chapter, take a look at the following resources:

- The Spruce Crafts has an in-depth guide to modeling with styrene at `https://www.thesprucecrafts.com/modeling-with-styrene-plastic-2382537`.
- *The Basics of Working with Styrene* at `https://www.youtube.com/watch?v=p3gabIJ3Ono`, by the Hawkins Screamer, also has great tips on building with styrene.
- *KiCad Like a Pro*, a Packt video course at `https://www.packtpub.com/product/kicad-like-a-pro-video/9781788629997`, shows you how to design circuits like the one we created in this chapter.
- A good guide for crimping is `https://mschoeffler.com/2017/11/02/tutorial-how-to-crimp-dupontmini-pv-connectors-engineer-pa-09-connector-pliers/`.
- The following Adafruit Collin's Lab crimping video may also be beneficial to you: `https://youtu.be/_zl28E2urEU`.

# 5
# Driving Motors with Raspberry Pi Pico

Our robot is looking ready to run. The first real test of a robot chassis is getting its motors to drive. This chapter will bring the robot to life, testing the wiring and motors, using **CircuitPython** on Raspberry Pi Pico. We will start with simple tests for each motor and then use them together to make movements. Finally, we will learn more sophisticated code to control their speed and end the chapter by making a path.

In this chapter, we're going to cover the following main topics:

- Driving forward and back
- Steering with two motors
- An introduction to PWM speed control
- Driving along a planned path

## Technical requirements

For this chapter, you will need the following:

- First is the built robot, as made in the previous chapters
- 6 x fresh AA batteries
- A PC or laptop with a USB micro cable
- Mu software to write our code and upload it
- Clear floor space with a meter or so in each direction to test the robot

All code examples are on GitHub at `https://github.com/PacktPublishing/Robotics-at-Home-with-Raspberry-Pi-Pico/tree/main/ch-05`.

# Driving forward and back

Our motors are attached, and the robot is looking ready to power up. First, we'll use CircuitPython to make test code to try each motor in turn. Then, when we have demonstrated the motors running, we'll make simple code to drive the motors straight forward and then back.

## Testing each motor with CircuitPython

We will start driving our robot by looking at how we connected our Raspberry Pi Pico to our motors in the following figure:

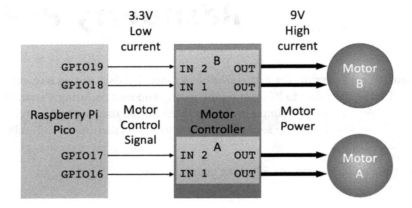

Figure 5.1 – Motor connections from Raspberry Pi Pico

*Figure 5.1* shows a closer look at the robot motor connections. On the left is Raspberry Pi Pico with four connections to the motor controller. They are on GPIO 16, 17, 18, and 19. These connections result in the motor controller powering the motor via one of the motor wires. Testing each of the Pico pins should cause a motor to do something.

Let's try this with some code, setting up one motor, and making it drive in a single direction. This example is called `motors_test_1_pin.py`:

```
import time
import board
import digitalio

motor_A1 = digitalio.DigitalInOut(board.GP17)
motor_A1.direction = digitalio.Direction.OUTPUT

motor_A1.value = True
```

```
time.sleep(0.3)
motor_A1.value = False
```

This code looks very similar to the LED code from *Chapter 2, Preparing Raspberry Pi Pico*. We set board.GP17, connected to the AIN1 pin on the motor controller, to a digital output.

The code sets GP17 to True, turning it on, then waits for 0.3 seconds, and sets the motor pin to False to turn it off.

Type this code into Mu and save it to CIRCUITPY. For this test, prop the robot up so that the wheels do not contact anything. To start this file, create a code.py module too:

```
import motors_test_1_pin
```

If you recall from *Chapter 2, Preparing Raspberry Pi Pico*, Raspberry Pi Pico will run the content of code.py when the robot starts or when you change a file on CIRCUITPY.

When you start this code, a motor moves for a short time and stops. The motor should run forward. If not, swap the motor output pins on this side. Repeat this exercise with GP18 and check whether the other motor runs forward too.

## How do you test this?

For all the remaining examples in the chapter, use the following procedure:

1. On the Pico (the CIRCUITPY drive), ensure the code.py file is empty; otherwise, it will rerun the old program until you update it.
2. Write your program and save it on your computer. You don't want to lose it!
3. Copy the program (and supporting files that have changed; we will be adding robot.py soon) onto the Pico.
4. Ensure the power switch is on and batteries are installed.
5. Have the robot propped up on something so that the wheels don't contact anything.
6. Update code.py to import the name of your program without the .py extension.
7. The first test of the code is seeing it run this way.
8. To run it for real, disconnect the Pico and turn off the power.
9. Then, put the robot in a clear space (the carpet or floor) and power it on.
10. Be prepared to pick it up and turn it off if it does something unexpected – the robot can damage itself if it drives into an obstacle without stopping.
11. You can also press *Ctrl + C* from the REPL at any time to stop a program running.

## Troubleshooting

There are some ways this could go wrong. Let's check a few here:

- If the import in code.py and your motors_test_1_pin filename do not match, you'll likely get an error in the REPL and nothing will happen.

- If the robot's UBEC is connected but the batteries are powered down, the robot may try to take too much power from the USB port. If you intend to test the motors, ensure the robot is powered on. If not, it may be good to disconnect the UBEC.

- If nothing still happens, disconnect the robot, turn it off, and check the wiring thoroughly – there should be no hot parts in the robot.

Now that you know how to test your code and make a simple example work, you are ready for more complicated examples.

## Testing multiple pins

Let's extend this code to test all the motor pins in motors_test_all_pins.py:

```python
import time
import board
import digitalio

motor_A1 = digitalio.DigitalInOut(board.GP17)
motor_A2 = digitalio.DigitalInOut(board.GP16)
motor_B1 = digitalio.DigitalInOut(board.GP18)
motor_B2 = digitalio.DigitalInOut(board.GP19)

motor_A1.direction = digitalio.Direction.OUTPUT
motor_A2.direction = digitalio.Direction.OUTPUT
motor_B1.direction = digitalio.Direction.OUTPUT
motor_B2.direction = digitalio.Direction.OUTPUT

motor_A1.value = True
time.sleep(0.3)
motor_A1.value = False
time.sleep(0.3)
motor_A2.value = True
time.sleep(0.3)
```

```
motor_A2.value = False
time.sleep(0.3)
motor_B1.value = True
time.sleep(0.3)
motor_B1.value = False
time.sleep(0.3)
motor_B2.value = True
time.sleep(0.3)
motor_B2.value = False
```

This code extends the first code example to test all the motor pins. Update code.py to import this instead, and you should see each wheel turn one way and then the other. If you do not see movement like this, please turn off the power/disconnect the robot and go back to *Chapter 4, Building a Robot around Pico*, to carefully check the wiring.

We will use these motors a lot. To save us from copying all the setup code each time, we can put it in a new file called robot.py:

```
import board
import digitalio
motor_A1 = digitalio.DigitalInOut(board.GP17)
motor_A2 = digitalio.DigitalInOut(board.GP16)
motor_B1 = digitalio.DigitalInOut(board.GP18)
motor_B2 = digitalio.DigitalInOut(board.GP19)


motor_A1.direction = digitalio.Direction.OUTPUT
motor_A2.direction = digitalio.Direction.OUTPUT
motor_B1.direction = digitalio.Direction.OUTPUT
motor_B2.direction = digitalio.Direction.OUTPUT
```

Now, you won't have to type that again. We can rewrite motors_test_1_pin.py:

```
import time
import robot
robot.motor_A1.value = True
time.sleep(0.3)
robot.motor_A1.value = False
```

You can run this (remember to change `code.py`) and apply the same change to `motors_test_all_pins.py`.

This code demonstrates that we can move the motors, but how do we use them together?

## Driving wheels in a straight line

In terms of motor direction, each motor pin controls this. Look at the following diagram:

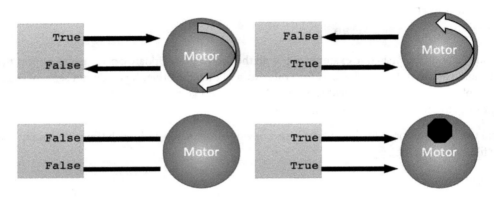

Figure 5.2 – Controlling motor direction

The preceding diagram shows a box representing the connection to the motor and its pins. Beside each is **True** or **False**, signifying the state of the controlling pin on a Raspberry Pi Pico. The arrows show the conventional current direction of power to the motor. With one pin high, current flows from this pin; if it is low, current flows back. When we set motor pins to opposite values, current will flow between the motor controller outputs and a motor turns. If the motor pins are both low, there is no current, and the motor is turned off and can coast. If they are both high, the motor will stop, like brakes.

> **Important note**
>
> In electronics, we describe current in two ways. **Conventional current** describes electricity flowing from the positive terminal of a power source to the negative terminal. However, the actual physics shows that negatively charged electrons flow the other way. We will stick to using conventional current to avoid confusion.

We can put this to use driving in a straight line. When we drive both motors together, a robot drives in a line forward or back. For example, look at the following figure:

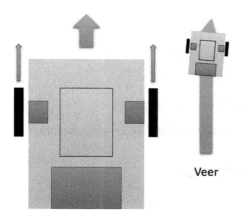

Figure 5.3 – Driving in a straight line

*Figure 5.3* shows a forward arrow from each wheel showing the motors are going forward. The combined wheel movement results in the robot driving forward – shown by the wide arrow in front of the robot.

The lines aren't quite straight as motors are slightly different, and it goes off slightly; we call this **veer**. We'll see how to correct this later in the book.

Let's make code to drive forward – `motors_forward.py`:

```
import time
import robot

robot.motor_A1.value = True
robot.motor_B1.value = True
time.sleep(0.3)
robot.motor_A1.value = False
robot.motor_B1.value = False
```

This code sets both motors forward by enabling (setting to `True`) the A1 and B1 pins. It waits for 0.3 seconds (300 ms) and then stops both motors. To drive for longer, you can increase the time. Using time to approximately control distance is not very accurate though.

Going backward means using the A2 and B2 pins instead (`motors_backward.py`):

```
import time
import robot

robot.motor_A2.value = True
```

```
robot.motor_B2.value = True
time.sleep(0.3)
robot.motor_A2.value = False
robot.motor_B2.value = False
```

Other than changing the pins, this code is identical.

We can now drive forward and backward, but how do we make turns?

## Steering with two motors

If we move one motor and not the other, the robot turns toward the wheel that isn't moving. For example, look at the following diagram:

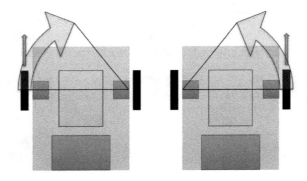

Figure 5.4 – Steering a robot with motors

*Figure 5.4* shows two turning robots. There is a forward arrow above the left wheel in the first panel, showing the wheel is driving forward. The right wheel is stopped. A transparent arrow superimposed on this shows the turn direction and that this turn pivots on the right wheel. The right robot shows an opposite turn. A robot can turn backward in the same way by reversing a single motor instead.

As we will do more with motors, we'll extend robot.py so that we can stop them all. Add this code at the end of robot.py:

```
def stop():
    motor_A1.value = False
    motor_A2.value = False
    motor_B1.value = False
    motor_B2.value = False
```

Ensure you copy this new version onto the Pico.

We can now use our turn principle in some code – `motors_1_motor_turn.py`:

```
import time
import robot

robot.motor_A1.value = True
time.sleep(0.3)
robot.stop()
```

This code example is very similar to the one-pin motor test. Only the `stop()` command is new. We roughly control the angle of turn by timing. It is tricky but possible to get 90-degree turns, but they won't be exact. Using different pins, we can turn using the motor on the other side, or reverse the current motor using the same principle.

What about using two motors? If we drive one motor forward and the other back, we can make a faster tighter turn and spin on the spot. Look at the following figure:

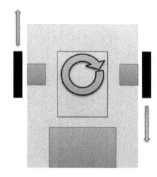

Figure 5.5 – Spinning with two motors

*Figure 5.5* shows the robot with an arrow going forward from one motor and back from the other. I've included a rotation arrow between the two wheels, showing the pivot for this turn. Let's see this in code; I suggest calling it `motors_2_motor_turn.py`:

```
import time
import robot

robot.motor_A1.value = True
robot.motor_B2.value = True
time.sleep(0.3)
robot.stop()
```

This code sets two pins high, A1 and B2 – you could say, diagonally opposite pairs. Driving A2 and B1 would spin the other way. Again, the timing controls the angle, and then we stop all the motors.

We can now move our motors, but they are on or off. However, we aren't controlling the speed, and we seem to be able to do one motor turn or full spin – what about gentler sweeping turns? The following section will get into pulse width modulation and how this controls motor speeds.

## An introduction to pulse width modulation speed control

**Pulse Width Modulation** (**PWM**) is how we control motor speeds from a digital control system. Instead of varying the voltage supplied to a motor, we use pulses to control it. The pulses are usually at a fixed rate, but the ratio of time-on to time-off changes. We call this the **duty cycle**. Controlling how much time per cycle the signal is on versus off will control the power getting to a motor. If the pulse is on for longer, the motor will go faster. The motor will go slower if the pulse is on for less time. So, at 50% time-on, the motor will be about 50% of its maximum speed.

The following diagram shows visual examples of this:

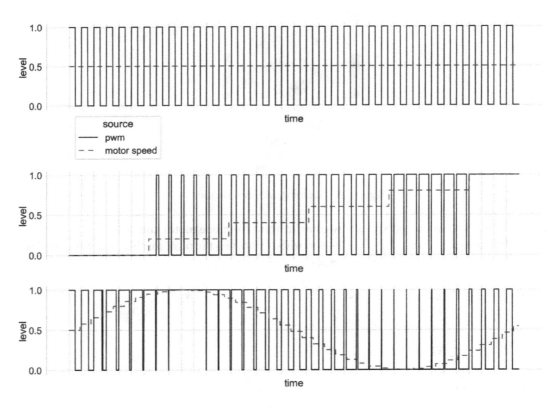

Figure 5.6 – PWM signals

The preceding diagram shows graphs of PWM signals. The top is a signal for driving a motor at half speed. The pulses on and off time are the same. The $X$ axis is the level, and the $Y$ axis is for time, with a solid line for the PWM signal and a dashed line for the power going to the motor. This panel shows a signal at half speed, with the duty cycle on for half the cycle.

The second graph shows control signals, ranging from 0 to completely on. When there are no pulses, this is equivalent to 0. When it is entirely on with no break, this is equivalent to 100%.

PWM can be fine-grained, with the bottom graph showing a sine wave along with the equivalent PWM signal for this. However, it's not smooth, as it can only change when there is a new pulse, and the levels also have a resolution.

## Driving fast and slow

We can take advantage of this PWM system to drive the robot at different speeds. Let's see how this works on a single wheel in `motors_pwm_drive_slower.py`:

```
import time
import board
import pwmio

A1_PWM = pwmio.PWMOut(board.GP17)

A1_PWM.duty_cycle = 2**16-1
time.sleep(0.3)
A1_PWM.duty_cycle = 2**15
time.sleep(0.3)
A1_PWM.duty_cycle = 0
```

In this example, we've gone back to setting up a pin from the board, but this time using `pwmio.PWMOut` instead of `DigitalIO`. We are using the motor A1 pin.

We then set `duty_cycle`, the amount of on-time to its highest value. It is a 16-bit value, so we use 2 to the power of 16 and subtract 1 – we don't need to do this calculation ourselves and can let the computer do it. This value will cause the motor to drive, as before, at full speed.

The code sleeps and then sets the duty cycle to 2 to the power of 15, half of our previous value. The motor will go at 50% speed here.

We let it run for 0.3 seconds and then set the `duty_cycle` to 0, which will turn off the motor. You can experiment with different values, but you may find the motor stalls (doesn't move) and beeps at values lower than half speed.

### Adjusting the robot library

We will need to apply changes to our robot library, where we set up motor pins as `DigitalIO` pins. We now need to use `PWMOut` to set the pins up. Here is a new `robot.py`:

```python
import board
import pwmio

motor_A1 = pwmio.PWMOut(board.GP17)
motor_A2 = pwmio.PWMOut(board.GP16)
motor_B1 = pwmio.PWMOut(board.GP18)
motor_B2 = pwmio.PWMOut(board.GP19)

def stop():
    motor_A1.duty_cycle = 0
    motor_A2.duty_cycle = 0
    motor_B1.duty_cycle = 0
    motor_B2.duty_cycle = 0
```

This code swaps the `Digital IO` setup we used before and uses `PWMOut` instead. The `stop` function now sets the pins `duty_cycle` to zero to stop all motor pins.

We can now use this to demonstrate the robot moving in `motors_convert_speed.py`:

```python
import time
import robot

max_speed = 2**16-1

robot.motor_A1.duty_cycle = int(0.8 * max_speed)
time.sleep(0.3)
robot.stop()
```

We use our refreshed `robot` class to set things up in this demonstration. First, we set a `max_speed` variable to hold the maximum value.

Having `max_speed` makes things more convenient, as we can then multiply it by a fraction between 0 and 1 to get a duty cycle value – here, we set the motor to 80% speed for 0.3 seconds and stop. We must use the `int()` function to convert the result to an integer (whole number).

We can move this multiplication up into the robot library. Add this code after the motor pins in robot.py:

```
max_speed = 2**16-1
right_motor = motor_A1, motor_A2
left_motor = motor_B1, motor_B2
```

First, there is max speed, as seen before, and then two variables to group our pins as respective motors. It means we can use pairs of pins, as shown next:

```
def set_speed(motor, speed):
    if speed < 0:
        direction = motor[1], motor[0]
        speed = -speed
    else:
        direction = motor
    speed = min(speed, 1) # limit to 1.0

    direction[0].duty_cycle = int(max_speed * speed)
    direction[1].duty_cycle = 0
```

This function will accept a motor (or pair of pins) as defined previously and then a speed between -1 (going in reverse) and 1 (going forward).

It checks whether the speed is negative and, if so, sets a direction variable with the motor pins swapped and makes the speed positive; otherwise, it just stores the current motor pins as a pair in direction.

The next line uses the min() function, which returns the minimum of two values. Putting in speed and 1 will limit the speed to no more than 1.0.

We then use the pins stored in the direction variable, setting the first pin's duty_cycle to the converted speed and the other pin to 0.

We can add two more functions on robot.py to make this more convenient:

```
def set_left(speed):
    set_speed(left_motor, speed)

def set_right(speed):
    set_speed(right_motor, speed)
```

These wrap the set_speed function, and they can now be used in your code with calls such as robot.set_left(1.0) and robot.set_right(0.8).

Let's use these and try a few different speeds in `motors_pwm_multispeed.py`:

```python
import time
import robot

try:
    for speed in range(5, 10):
        robot.set_left(speed/10)
        robot.set_right(speed/10)
        time.sleep(0.3)
finally:
    robot.stop()
```

The previous example uses a `for` loop, looping over the numbers 5 to 10. This means we get a `speed` variable in each loop. We divide this by 10, so we now get 0.5 to 1.0 and use the `robot.set_...` methods to set both motors to this `speed`. The program then sleeps for 0.3 seconds and loops to the next item.

It is wrapped in `try...finally` so that the program will always call `robot.stop()`, even if something fails in our loop; this ensures that the motors don't keep driving.

The robot will start slowly and then speed up and stop when you run this example.

We can use the variable motor speeds to make gentle sweeping turns too. Let's see how in the next section.

## Turning while moving

We can make gentle, sweeping turns by sending different speeds to each motor using PWM. For example, look at the following figure:

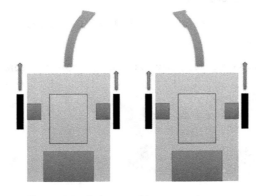

Figure 5.7 – Robots making sweeping turns

*Figure 5.7* shows robots making gentle sweeping turns to the right and the left. The speed difference between the motor's controls where the turn's pivot is. The closer the speeds are, the wider the turn radius is. This is demonstrated in `motors_pwm_gentle_turn.py`:

```
import time
import robot

try:
    robot.set_left(1.0)
    robot.set_right(0.5)
    time.sleep(1)
finally:
    robot.stop()
```

This example uses the same `try...finally` construct as before. It sets the left motor to full speed and the right to half speed, making a wide arc to the right for half a second. The robot then stops.

We've now seen how to control a robot's motor speeds with PWM and make different turns. Finally, we are ready to put some of this together to make the robot drive in a pre-determined path on the floor.

# Driving along a planned path

We can use our straight-line driving motions and curved turns to make an almost square path on the floor. We can use the helper functions we've made to keep this short.

## Putting line and turn moves together

We are going to put some of our learning together to make a simple square pattern, as the following diagram shows:

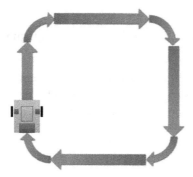

Figure 5.8 – Driving a square path

The figure shows a square made up of four straight lines and four turns. These eight instructions are four repeating sets of a straight line and then a turn. We will have to adjust the timing of the turn to make it close to 90 degrees.

We start this code with some helpers for our motions – in `pwm_drive_square.py`:

```python
import time
import robot

def straight(speed, duration):
    robot.set_left(speed)
    robot.set_right(speed)
    time.sleep(duration)

def left(speed, duration):
    robot.set_left(0)
    robot.set_right(speed)
    time.sleep(duration)
```

The `straight()` function just puts both motors going forward (or back) for the duration. The `left()` function stops one motor and drives the other at speed.

We can then use these in a main `for` loop to get the four turns:

```python
try:
    for n in range(0, 4):
        straight(0.6, 1.0)
        left(0.6, 1.0)
finally:
    robot.stop()
```

Our loop counts four times using the `range()` function. We then use the `straight` and `left` functions with speed and duration.

Note that the performance of this will vary greatly, depending on how fresh the batteries are and which surface you drive on – friction will slow motors down. You will likely need to adjust the relative times of the `straight` and `left` function uses to try and get a square.

## The flaw with driving this way

You'll notice that the path we planned is not quite the plan we got. Even after some time adjusting it, you probably got a path like the following:

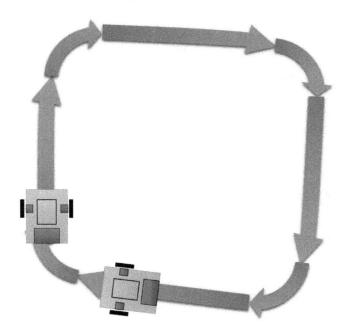

Figure 5.9 – What your path might look like

*Figure 5.9* shows an exaggerated version of what you may have got. Even after much tweaking, the robot may have made turns slightly above or below 90 degrees. The robot may also have veered slightly while making the straight lines. In the next chapter, we will pick up our first sensor, the encoders, which can be used for more accurate turning and to correct for veer.

The moment you change surfaces or the batteries degrade, the robot will go off course further. Drawing a path like this is a start, but later in the book, we'll see how to use sensors to improve the accuracy of such motion.

## Summary

In this chapter, we learned how a controller such as Raspberry Pi Pico uses a motor controller to drive motors. We saw how to control motor directions to drive in a straight(ish) line or make a robot turn.

We then learned about PWM control and how to vary motor speeds, creating a handy `robot` Python library for our robot in the process. Next, we used this to make sweeping curves and build a multi-step path example with our robot. This path code did, however, show up some accuracy shortcomings.

We have used time to estimate how much we move our motors. However, motors don't start immediately, and they can sometimes stick. In addition, some motors are slower than others. Therefore, we are working only with approximates. In the next chapter, we will look at how to measure how much the motors have turned to get a more accurate estimation of the robot's movement.

## Exercises

Now you've had a taste of driving the robot motors, perhaps you can practice your skills with the following challenges:

- Can you make other shapes with this method, such as a triangle, or, by using gentle turns, drive in a circle?
- What is the lowest PWM value before the robot stalls on two motors?
- Does the preceding value change on one motor?
- How does the robot's driving behave on a different surface, such as carpet or wood?

## Further reading

These additional resources will help you learn more about the concepts covered in this chapter:

- A great YouTube video by Afrotechmods shows more about PWM: https://www.youtube.com/watch?v=YmPziPfaByw.
- Sparkfun has an explanation of PWM at https://learn.sparkfun.com/tutorials/pulse-width-modulation/all where, in addition to its use of motors and servo motors, it shows how you can use the technique to control LED brightness.
- Adafruit, the team that created CircuitPython, has a PWM programming tutorial at https://learn.adafruit.com/circuitpython-essentials/circuitpython-pwm.
- My book *Learn Robotics Programming, Second Edition*, also published by Packt, has a chapter covering motor steering mechanisms, including the one seen here, with Python code for Raspberry Pi.

# Part 2: Interfacing Raspberry Pi Pico with Simple Sensors and Outputs

In this part, we will build upon the basic Raspberry Pi Pico knowledge, adding more complicated devices and code to interface them. We will see how sensors can interface our robot more with the real world. We will add Bluetooth LE to link with a computer.

This part contains the following chapters:

- *Chapter 6, Measuring Movement with Encoders on Raspberry Pi Pico*
- *Chapter 7, Planning and Shopping for More Devices*
- *Chapter 8, Sensing Distances to Detect Objects with Pico*
- *Chapter 9, Teleoperating a Raspberry Pi Pico Robot with Bluetooth LE*

# Measuring Movement with Encoders on Raspberry Pi Pico

Robots don't just run code blindly; they need sensors. What sensors do we add first? Our robot already has some sensors on board, and in this chapter, we'll see how to use them.

We finished the previous chapter noticing that timing isn't the most accurate way to determine robot movement. So, let's see how our first sensor, encoders, can improve this. Encoders are the first step in getting accurate movement and location estimation in robots. We will also learn one of Raspberry Pi Pico's excellent features – **Programmable IO (PIO)**.

We'll dig into movement fundamentals, odometry and encoding, look at Raspberry Pi Pico PIO in CircuitPython, and use this to get data from the encoders on our robot.

In this chapter, we're going to cover the following main topics:

- About encoders and odometry
- Wiring in encoders on a Raspberry Pi Pico robot
- Programming Raspberry Pi Pico PIO
- Measuring encoder count for movement

## Technical requirements

For this chapter, you will need the following:

- First is the built robot, as made in the previous chapters
- 6 x fresh AA batteries
- A PC or laptop with a USB micro cable

- Mu software to write our code and upload it
- Clear floor space with a meter or so in each direction to test the robot

All code examples are on GitHub at `https://github.com/PacktPublishing/Robotics-at-Home-with-Raspberry-Pi-Pico/tree/main/ch-06`.

# About encoders and odometry

**Odometry** is measuring how a position has changed over time. We can combine measuring and estimation to determine where you are on a route. **An encoder** is a sensor designed to measure distance traveled via wheel turns. They are like **tachometers**, but encoders measure position whereas tachometers measure only speed. Combined with time, they can make a speed measurement too.

## Absolute and relative sensing

Sensors for a robot's location come in two primary forms. They are as follows:

- **Absolute sensors** encode a position to a repeatable position. They have a limited range or resolution, such as encoding a position along a known line. For example, GPS sensors have exact positioning with low resolution, suitable for meters but not millimeters.
- **Relative sensors** tend to be cheaper. However, they produce a relative change in position, which needs to be combined with the previous state to get an absolute estimate – this means that errors can accumulate. Relative encoders are one example of relative sensors, also known as **incremental encoders**.

If a sensor tells you where something is *at*, it is absolute. It is relative if it tells you how much something has moved *by*.

## Types of encoders

Most encoder designs work by passing markers or code over a sensor that counts or decodes the pulses.

Encoders come in a few forms. Some example types are as follows:

- A potentiometer or **variable resistor** can sense an absolute encoder position by measuring resistance. Servo motors use them. However, potentiometers are not suitable for continuous rotations such as wheels, as their track lengths limit them. In addition, regular movement wears them down, as they move contacts across each other.
- **Mechanical encoders** pass electrical contacts over each other, producing on and off pulses. They are subject to heavy wear, so I do not recommend them.

- **Optical encoders** shine a light through a disk or strip with slots and detect the passing of slots in front of light sensors. They can come in absolute and relative flavors. They can be susceptible to interference from light sources or just dirt.

- **Magnetic encoders** detect the movements of magnets in a disc using **hall-effect sensors**. Dirt, light interference, and physical wear do not affect them so much. Hall-effect sensors produce a voltage depending on a magnetic field – encoder modules produce pulses from this.

The motors we chose came with rotary magnetic encoders in a convenient and small format as part of the package. They are incremental encoders.

## Encoder pulse data

We can better understand encoders by looking at the pulses they output. Relative encoders usually output digital pulse chains, 1s and 0s. The simplest form is just to count pulse edges by detecting marks passing a sensor, as the following diagram shows:

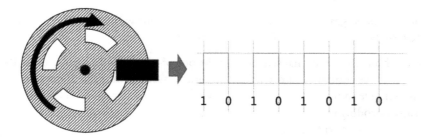

Figure 6.1 – Simple pulse encoding

On the left of *Figure 6.1* is a disk with a rotation arrow. On the disk are four white marks – representing the markers passing a sensor. The black rectangular object is the sensor that detects the markers. The sensor produces a value of 0 when it isn't detecting a marker and 1 when it is – making a bunch of pulses or a stream of binary bits with values of 1 or 0.

We can count the pulses to get an idea of far the wheel has turned. We count a high pulse as 1 and a low pulse as 0. We can also count **edges**, the changes from 0 to 1 and 1 to 0. Counting edges gives us eight steps per wheel turn. The graph to the right of the following diagram shows these pulses.

We are likely to want to increase that sensitivity and detect which direction a wheel is going. To do that, we add a second sensor to the same wheel, as the following diagram shows:

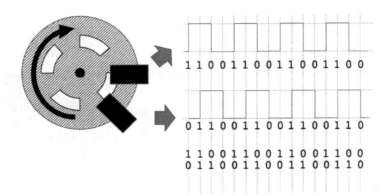

Figure 6.2 – Quadrature pulse encoding

*Figure 6.2* adds a second sensor to catch the markers at a slightly different time from the first, generating pulses out of phase; this means we have two streams of pulses.

At the top right is the pulse graph for the first sensor, with the digits we had read from times between the pulses added, showing a pulse train of 16 bits for the same period.

The middle graph shows the additional sensor, with the out-of-phase 16-digit pulse train. Below the graphs are the two states combined at each point in time as **Gray code**, with 2 bits of information on where we are relative to the last position. We have twice as many edges, increasing the sensor resolution and also encoding the wheel's direction. If we reverse the wheel, this sequence will reverse. This system is known as **quadrature encoding**.

Let's look at the encoders on our robot in the next section.

# Wiring in encoders on a Raspberry Pi Pico robot

Our robot has already got encoders on board, and we have already wired them in. We can take a closer look at the motors and how they are wired into Raspberry Pi Pico GPIO pins to understand the robot better.

## Examining the motors

We use N20 geared motors with encoders. The following diagram labels the motor parts:

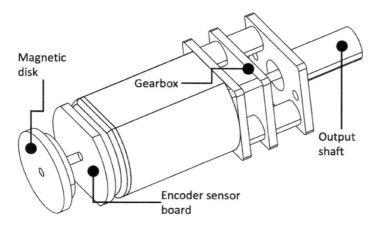

Figure 6.3 – The N20 motor parts

*Figure 6.3* shows a drawing of the motors we have used. Marked on it are essential features that affect how we use the encoders. On the left is a magnetic disk with markers in it. This disk is attached to the motor's driveshaft and sensed by the encoder sensor board. On the right are the gearbox and the motor output shaft.

The driveshaft goes through the gearbox, so the output shaft will not make the same number of rotations as the disk – the gear ratio will determine this relationship. So one revolution of the output wheel could count many pulses; this gives us high resolution.

Consult the datasheet for the motors. Some Chinese characters are likely, but important numbers are usually in English. You may need translation services built into web search engines here. The datasheet and product page have two important numbers, the number of encoder counts per disk revolution and the gear ratio. The datasheet may note counts per disk revolution as pole count.

In my case, the gear ratio is 298:1, and the pole count is 14. Interpreting these facts means I get 298 turns of my encoder wheel per output wheel revolution. Each encoder turn produces 14 poles on each sensor (two sensors), so we get 28 edges. Multiplying the number of sensor pulses by the gear ratio gives 8344 edges per turn.

## Examining the wiring

We saw the wiring for our robot in *Figure 4.20* of *Chapter 4, Building a Robot around Pico*. However, to better illustrate the encoder connections, here is a diagram focusing only on the wiring of encoders to Pico:

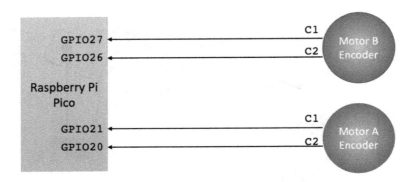

Figure 6.4 – Encoders wired to Raspberry Pi Pico

The preceding figure takes a closer look at data connections for a robot encoder connection schematic. On the left is Raspberry Pi Pico; this has four connections from the encoders. These are on GPIO 20, 21, 26, and 27. Each of these can be set as input pins to read the state of the encoder pins.

If we were just reading encoders alone, we could write code to check each pin in sequence. However, doing this may tie things up. What if we could get components of the Pico to monitor these pins and pulse chains for us so that we could just read a counter for them when we need it?

## Programming Raspberry Pi Pico PIO

We saw the PIO system back in *Chapter 1*. We could read encoders in Python on the Pico central cores; however, we can make monitoring the encoders the responsibility of PIO, letting those central cores do other things. The PIO system can read and decode the Gray code emitted by the encoders. This section will show how to program PIO in assembler and load the programs with CircuitPython.

### Introduction to PIO programming

As we saw in *Chapter 1*, Pico has two PIO devices, each with four state machines. We program PIO in **assembly language**. Instructions command PIO to perform operations such as manipulating IO pins, registers, and **first in first out** (**FIFO**) queues. The following diagram is a simplified representation of a state machine:

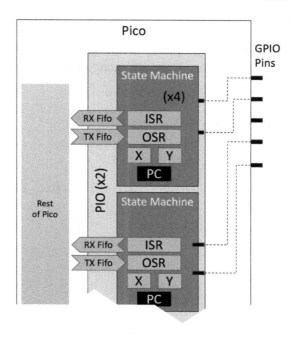

Figure 6.5 – The Raspberry Pi Pico PIO state machines

The preceding diagram shows Raspberry Pi Pico PIO state machines, highlighting registers and other features. It shows the two PIO devices and the state machines in them.

A **register** is like a variable; they have fixed names, and there are only a few per core. PIO registers are 32-bit and specific to each state machine. We use an essential subset: **input shift register** (**ISR**), **output shift register** (**OSR**), and $X$ and $Y$ (scratch registers). They can store a number or a binary pattern. It is common to refer to 32 bits as **words**.

Each state machine runs the program independently, so the same code runs four times, with independent registers. In addition, mappings (shown as dashed lines) connect state machines to IO pins – the tiny rectangles. Code can configure which state machines use which pins and a single state machine can read/write from many pins.

State machines also have FIFO queues. Data put into a FIFO queue comes out in the same order. Each can hold 4 x 32-bit words. These let PIO **transmit** (**TX**) data from or **receive** (**RX**) data to other devices within the rest of Pico. We can configure FIFO queues in many ways. For this chapter, we will use the RX FIFO queue to receive data from PIO to our code.

Each PIO block can run an independent program with 32 instructions – each roughly, but not quite, a line of code. But how do we write and use these programs?

## Introducing PIOASM

In CircuitPython, we assemble the PIO code using the Adafruit CircuitPython `PIOASM` library. This runs on Pico, taking the assembly code in a string and outputting a sequence of bytes with the code in it.

We need to put this `PIOASM` library onto Raspberry Pi Pico:

1. First, find the `CIRCUITPY` volume on your computer.
2. From the Adafruit CircuitPython library (as discussed in *Chapter 2*), copy `lib/adafruit_pioasm.mpy` into the `lib` folder on `CIRCUITPY`.

With that in place, we can write our first PIO-based program. Here's `pio_led_test.py`:

```python
import time
import board
import rp2pio
import adafruit_pioasm
```

The first four lines are imports. You've seen `time` and `board` before. The `rp2pio` library lets us communicate with the PIO blocks and start code and interact with state machines. The `adafruit_pioasm` library turns assembly code into bytes that PIO state machines can run. Now, let's get into the assembly code:

```python
led_flash = """
    pull
    out pins, 1
"""
```

The preceding code puts some PIO assembly into the `led_flash` string. Triple quotes in Python declare a long multi-line string.

The assembly code starts with the `pull` instruction; this gets a word from the TX FIFO queue (removing it) and stores it in the OSR. The `out pins, 1` instruction puts 1 bit of data from the OSR onto the configured pins – setting the state of a single pin. The code wraps around to run at the first instruction in a loop. We need to assemble this code:

```python
assembled = adafruit_pioasm.assemble(led_flash)
```

The `adafruit_pioasm.assemble` function generates bytecode, which we store in the `assembled` variable. We can run this:

```python
sm = rp2pio.StateMachine(
    assembled,
```

```
    frequency=2000,
    first_out_pin=board.LED,
)
```

`rp2pio.StateMachine` requests a state machine to run this code, telling it how fast to run and which output pin to map to – in this case, `board.LED`. Our code will be running on the PIO, but we have no data in the FIFO queue, so it will be waiting for us to write something to it.

We can write data with a loop in this program:

```
while True:
    sm.write(bytes([1]))
    time.sleep(0.5)
    sm.write(bytes([0]))
    time.sleep(0.5)
```

This loop writes 1s and 0s to the state machine (in the `sm` variable). It must wrap the data as a list since a FIFO queue can store more than one data element as a `bytes` type.

Send this to Raspberry Pi Pico, and the LED will flash. Let's ensure this works.

### Troubleshooting PIO code

Writing assembler code is somewhat tricky the first time – these tips can help you get moving:

- This code uses the Python triple quote, `"""`, for a multi-line string. Ensure you have three quotes at both ends of the assembly section, or you will see errors.
- If Pico cannot load `adafruit_pioasm`, ensure you have followed the setup steps to copy the mpy file into the `lib` folder on Pico.
- Note that there must be two close brackets after the `sm.write` statements.
- If Pico is not running your code, remember to import your code in the `code.py` file.

These tips should get you up and running.

Now we have our first PIO code, we can try reading data back from an I/O pin.

## Detecting input with PIO

Fetching input from PIO is as simple as getting a pin state into a register and pushing that onto the RX FIFO queue for the Python code to pick up. Create a file called `pio_read_1_pin.py`. We will add one more import to read PIO data:

```
import board
import time
```

```
import rp2pio
import adafruit_pioasm
import array
```

The assembly section looks like this:

```
pio_input = """
.program pio_input
    in pins, 1        ; read in pin (into ISR)
    push noblock      ; put this into input FIFO
"""

assembled = adafruit_pioasm.assemble(pio_input)

sm = rp2pio.StateMachine(
    assembled,
    frequency=2000,
    first_in_pin=board.GP20
)
```

The `in pins, 1` instruction will read 1 bit of data from 1 input pin and store this in the ISR. Following this is a comment starting with a `;` character that is for humans to read. The spaces are optional and are there to aid readability by aligning the comments. You can also add a `.program` line – effectively another comment.

The next instruction is `push noblock`, which will take the ISR register and push it as a word onto the RX FIFO queue. `noblock` ensures it will not wait for the FIFO queue to be empty – note that data is not written to the FIFO queue if it is full.

We then assemble this code and load it into a state machine, passing `first_in_pin` to map one of our encoder pins as input. Next, we need a buffer to read our FIFO queue data where the Python code can use it:

```
buffer = array.array('I', [0])
```

The array type makes fixed-size data structures in memory. It specifies an unsigned 32-bit integer with `'I'`. We size it as 1 element and initialize it as `0`.

The `main` loop reads data into the buffer and prints it:

```
while True:
    sm.readinto(buffer)
```

```
    print(f"{buffer[0]:032b}")
    time.sleep(0.1)
```

The `sm.readinto` Python function pulls data from a FIFO queue into a waiting buffer. It will wait if there is no new data to fetch.

We then use a fancy print to output our data. Python f-strings (prefixed with an f) let you use a variable in the string – in this case, extracting the only element of `buffer`. The `:032b` format specifier tells Python to format the data as 32-digit binary, with the empty digits in front filled with 0s.

When you run this, you will see a repeating output with one of two states:

- 10000000000000000000000000000000, showing the encoder pin is high
- 00000000000000000000000000000000, showing the encoder pin is low

Turn the wheels on the robot slowly. One of them will make the pin change states. It may be surprising that the 1 bit is at the start of the data (and would be there in the ISR before we sent it).

We can extend this code to work with two pins easily. Copy this to `pio_read_2_pins.py` and make the following modification:

```
pio_input = """
.program pio_input
    in pins, 2       ; read in two pins (into ISR)
    push noblock     ; put ISR into input FIFO
"""
```

The other code remains the same, except that when we run it to turn the wheel slowly, the output will now show 2 bits from the encoder, in the following four states:

```
11000000000000000000000000000000
01000000000000000000000000000000
10000000000000000000000000000000
00000000000000000000000000000000
```

These are the bits of the quadrature encoding discussed previously!

### Troubleshooting

This section is the first time we have tried to get information from the encoders, and issues may occur. Try these steps:

1. If the data values are not changing, check the wiring carefully.
2. If only 1 bit is changing, 1 encoder wire may be incorrect.

We have data, but the PIO can work harder to decode this and count it for us. Next, let's look at these PIO instructions and how they interact with registers.

## PIO instructions and registers

Understanding how registers are changed and manipulated by PIO instructions is crucial to writing and understanding PIO code. The PIO has nine instruction types; however, they have several modes, making their use nuanced and complex. *Chapter 3* of the *RP2040 datasheet* from Raspberry Pi serves as a comprehensive reference. We can get familiar with a few more here.

### Debugging a register

This next example shows how to store a value in a register and surface it to code for printing. We keep the imports as before. Call this file pio_debugging_registers.py:

```
program = """
    set y, 21
    mov isr, y
    push noblock
"""


assembled = adafruit_pioasm.assemble(program)
```

This code uses the set instruction, which can put any value less than 32 (5 bits) into a register. In the example, we store it in the y register.

The following line is mov isr, y, which copies the right (y) register data into the ISR. We must store a value in the **ISR** to use it in a push statement, putting it into a FIFO queue.

We assemble this and send it to the state machine:

```
sm = rp2pio.StateMachine(assembled, frequency=2000)
```

We can then pull this data from the FIFO queue and examine the content as decimal and binary:

```
buffer = array.array('I', [0])
sm.readinto(buffer)
print("{0} 0b{0:032b}".format(buffer[0]))
```

This code will run and simply pass the number 21 through the system. Because I know the value is low, I have formatted it with 32 leading 0s. Therefore, the output should look like this:

```
code.py output:
21 0b00000000000000000000000000010101

Code done running.
```

This feature is handy, and we can use this code with different assemblers as a template to test out PIO assembly techniques. We can start by looking at manipulating bits in registers.

### Bit manipulations

When dealing with registers, we may want to manipulate their content. For example, we may want to move bits around, shift them, reverse their order, or flip them wholesale. The following diagram shows these operations:

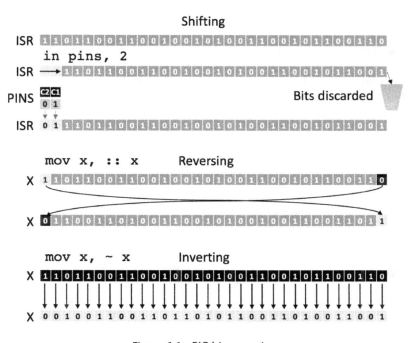

Figure 6.6 – PIO bit operations

*Figure 6.6* shows these frequent operations pictorially. The tables show the state of registers as bits, with the register name on the left. The diagram shows assembler instructions with their effect on the bits highlighted.

We have already talked a lot about bit shifting into the ISR. When we use the `in` instruction, it makes space for the number of bits to shift in, discarding bits at the end. It then copies the new bits into the space. Code can alter this shift direction with `StateMachine` parameters, but the operation is essentially the same. The example for shifting uses `in` to read the C1 and C2 pins from our encoder. However, as well as pins, the source can be other registers or null to copy in zeros. This operation does not alter the source.

We can **reverse** the content of a register using the `::` operation in `mov`; this can be useful to get to bits at the other end. The following assembler demonstrates this:

```
set y, 21
mov isr, :: y
push noblock
```

You can use this example in the same Python code shown in the preceding debug registers example. Note that we reverse the `y` register right into the ISR. As the code gets more complicated, combining operations like this will be critical, as with only 32 instructions, every instruction counts! The output of that code looks like this:

```
2818572288  0b10101000000000000000000000000000
```

**Inverting** a register is a bitwise `not` operation. It replaces every bit with its logical opposite – 1s become 0s and 0s become 1s. If we represent signed numbers with our 32-bit words, this will make them negative. Try this assembler code:

```
set y, 21
mov isr, ~ y
push noblock
```

This code is exactly like the preceding example, producing the inverted result:

```
4294967274  0b11111111111111111111111111101010
```

We can modify our buffer to see how a value becomes negative too. Change the array format to a lowercase `i`:

```
buffer = array.array('i', [0])
```

When running this, we can see what the output is like:

```
-22 0b-00000000000000000000000000010110
```

The binary makes less sense, but we can see that the decimal number is the negative plus one. Inverting again gets back to the original number.

### Extracting the value of a single bit

One more thing to do is to combine operations so that we can extract a specific bit using a couple of bit shifts. The following diagram shows the operation:

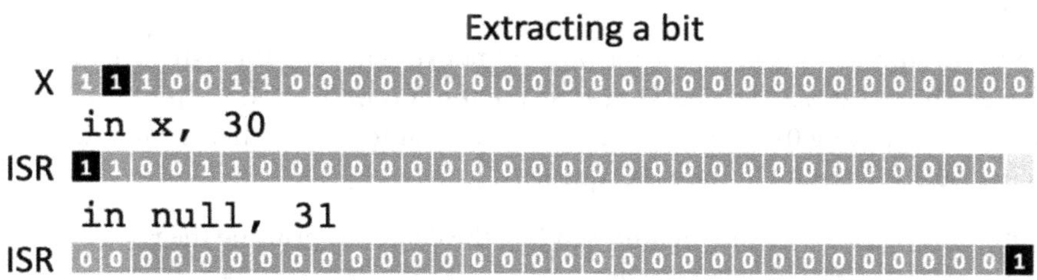

Figure 6.7 – Extracting a bit

Let us imagine our value starts in x; I've shaded the bit we want darker than the rest. The assembly snippet for this is the following:

```
in x, 30
in null, 31
```

The first instruction shifts the bits up to the one we want into the ISR. The diagram shows that the bit we want is now leftmost in the ISR. However, as we do not want anything to the right of this bit, we use the in instruction to shift in 31 zeros from null. This shift pushes all the other bits away – so we have only 0s and the bit we want.

We'll put this to real use with the OSR instead. Put the following code in extract_a_bit.py. We start with imports and a variable to tweak the behavior:

```
import rp2pio
import adafruit_pioasm
import array


bit_to_extract = 30
```

Remember that we store the assembler code in a Python string, so we can perform Python string formatting on it with an f-string:

```
program = f"""
    pull block
    in osr, {bit_to_extract}
    in null, 31
```

```
    push noblock
"""
```

Since we cannot use the `set` instruction with values higher than 5 bits, we start by pulling data to read, which goes into the OSR.

We then use `in` to shift the bits up to the bit we want to extract, using the f-string to substitute the variable here. The bit to extract must be 32 or less. The result is in the ISR. We perform a further `in` operation from `null`, using 0s to drop all but the bit we wanted. Since this result is already on the ISR, we can use `push` to send it to the FIFO queue.

The remaining code handles assembling this, sending the data, and printing the result:

```
assembled = adafruit_pioasm.assemble(program)
sm = rp2pio.StateMachine(assembled, frequency=2000)

sm.write(array.array('I',
  [0b01101000_00000000_00000000_00000000]))

buffer = array.array('I', [0])
sm.readinto(buffer)
print("{0} 0b{0:032b}".format(buffer[0]))
```

The result from running this should be the following:

```
1 0b00000000000000000000000000000001
```

If we want to use this data for conditional logic, we could use the `jmp` instruction, requiring us to use the `mov` instruction to move the data into `x` or `y` scratch registers.

## Making a counter with PIO

Counting requires us to be able to add or subtract from a register. We'll start with counting down, as that is easier.

At first glance, the datasheet shows no arithmetic instructions in the PIO instruction set. So how do we add or subtract? Although we have not used it yet, the `jmp` instruction in PIO assembler usually jumps to a label somewhere else in our assembler instructions. However, it has a trick – it can subtract from a scratch register. We can exploit this side effect to perform simple arithmetic for us. Using `pio_debugging_registers.py` as a template, try switching out the assembler for the following code:

```
    set y, 21
    jmp y--, fake
```

```
fake:
    mov isr, y
    push noblock
```

I put this with the template in `pio_counting_down.py`.

Sending it to Pico and running it gives the following output:

```
20 0b00010100
```

Hurrah – we can subtract! We have used a `fake:` label for the subtraction because we don't want to jump anywhere, just perform the arithmetic.

What about adding? This is trickier, but if you recall the bit invert, the number flipped from positive to negative. We can exploit this by subtracting 1 and flipping it again.

Use this assembler (in `pio_counting_up.py`):

```
    set y, 21
    mov y, ~ y
    jmp y--, fake
fake:
    mov isr, ~ y
    push noblock
```

We still have our fake label, but we flip the y value into itself first and then flip it again when putting it into the ISR. The output of running this is as follows:

```
22 0b00010110
```

You have seen techniques for working with the PIO, how to read data from pins, extract information from it, and perform arithmetic. We have building blocks. The following section will see us use them to decode information from the encoder pins into a counter.

# Measuring encoder count for movement

We know what sequences to expect for our encoder, and we have a working knowledge of PIO assembler. So, we can bring these together to create the counter. We'll start simple though; let's see how to detect when a system has changed.

## Making a simple PIO change detection loop

As we saw in the read two pins example, when we output the system's state in a tight loop, it floods off anything interesting. We are interested in state changes, a step toward the full decoder. In the `pio_one_encoder_when_changed.py` file, we go straight from imports into the assembler:

```
import board
import rp2pio
import adafruit_pioasm
import array
```

We start by clearing y  -; we are going to use y to store a pin value for comparison:

```
program = """
    set y, 0
```

The following code creates a `read` label; we can loop to this point to get new pin readings. It stores the old y in x so that we can get a new value. Shifting `in null, 32` will fill the ISR with zeros, clearing it. We can then get two pins in the ISR:

```
read:
    mov x, y
    in null, 32
    in pins, 2
```

We want to compare our new value in the ISR with our old value, now in x. However, `jmp` cannot use the ISR for comparisons, so we first copy the ISR into y.

```
    mov y, isr
    jmp x!=y different
    jmp read
```

As the code shows, we can now use `jmp x!=y` to jump somewhere else when the register values are different – to a label named `different`. If we do not find them different, we loop back to `read` to try a fresh sample from the pins with the unconditional `jmp`.

Let's see the code in the `different` label:

```
different:
    push noblock
    jmp read
"""
```

Although we copied the ISR to y, it is still in the ISR so that we can push this value, the new changed value, out to the Python code and then jump back around to read it again. So, the total effect is that it will spin reading values and, if they are different, push them to the Python code and then go back to spinning in the read loop.

Let us continue the Python code:

```
assembled = adafruit_pioasm.assemble(program)
sm = rp2pio.StateMachine(
    assembled,
    frequency=20000,
    first_in_pin=board.GP20,
    in_pin_count=2
)


buffer = array.array('I', [0])
while True:
    sm.readinto(buffer)
    print("{:032b}".format(buffer[0]))
```

The last half assembles our code and creates a state machine with it, using a higher frequency. It sets the state machine to use GP20 as the first input pin. Then, it uses in_pin_count to set a range of two input pins, matching one of the encoders. It then reads data into a buffer and prints it in a loop. The sm.readinto method waits until there is data, so the Python code only prints when there is a change. Try rotating the wheels slowly, and you should see the output change:

```
00000000000000000000000000000000
01000000000000000000000000000000
11000000000000000000000000000000
10000000000000000000000000000000
00000000000000000000000000000000
01000000000000000000000000000000
11000000000000000000000000000000
10000000000000000000000000000000
00000000000000000000000000000000
```

We can see the encoder output, and only when it changes. We could just count the changes, but our system should count in different directions depending on the wheel movement. Let's write code to check this.

## Making a bidirectional counter with PIO

We can detect when our sensor is in a new state and store that in y, with an old state in x for comparison. We also need to store a counter, and since we aren't using it, the OSR will suffice. We'll jump right into the assembler, since the imports don't change:

```
program = """
    set y, 0                ; clear y
    mov osr, y              ; and clear osr
read:
    mov x, y
    in null, 32
    in pins, 2
    mov y, isr
    jmp x!=y, different
    jmp read
```

As you can see, beyond setting up the OSR, this starts the same as the previous example. However, where things are different, we need to be more innovative. Comparing the 2 bits with the previous 2 bits is tricky in assembler, and we have a 32-instruction limit. What we are trying to evaluate is the sequences in the following diagram:

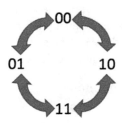

Figure 6.8 – Quadrature encoding sequence

*Figure 6.8* shows the sequence for encoder signals. Each pair of numbers shows the sensor states. A jump from **00** to **10** suggests the encoder is going clockwise, and from **00** to **01** is anticlockwise; we can follow the sequence around in either direction.

We can evaluate an old reading and a new reading with the following conditions:

- If the first bit of the old reading is 0

  - And the current second bit is 1

    - Then it is going anticlockwise

- Else, the current second bit is 0

    - It is going clockwise

- If the first bit of the old reading is 1

    - And the current second bit is 1

        - It is going clockwise

    - Else, the current second bit is 0

        - It is going anticlockwise

This logic can boil down to a few instructions in the assembler. First, we can isolate the bit we want in the x register (the old reading) and compare that to zero, jumping on the result:

```
different:
    in x, 31
    in null, 31
    mov x, isr
    jmp !x, c1_old_zero
```

Note that this uses the preceding bit extraction method; bit 31 would be the first pin (C1) read in. We now have the old C1 value in x, padded with 0s. If the x register is zero, the jmp !x instruction will jump to the c1_old_zero label. Otherwise, it will fall through.

For clarity, we will start the fall-through section with the c1_old_non_zero label; this is just a comment, though:

```
c1_old_not_zero:
    jmp pin, count_up
    jmp count_down
```

At this point, we test a pin. We'll see later that we can set jmp_pin when creating the state machine, and we'll set it to the C2 pin for an encoder, so this will have the current second pin in it. The jmp pin, count_up instruction will jump to the count_up label if the current state of the pin is 1. Otherwise, we unconditionally jump to count_down.

The code for when C1 is zero is the opposite:

```
c1_old_zero:
    jmp pin, count_down
    ; fall through
```

However, I am being sneaky – instead of the unconditional jump, the next code section will be count_up, so we can skip it and save an instruction. count_up is adding 1, as we've seen before, with the twist that the original value is on the OSR:

```
count_up:
    mov x, ~ osr
    jmp x--, fake
fake:
    mov x, ~ x
    jmp send
```

We invert the OSR into x, use jmp to jump to a fake label, subtract 1 from x, and then invert x back. Finally, this code jumps to send the data, with the new count now in x.

count_down is a little simpler:

```
count_down:
    mov x, osr
    jmp x--, send
```

This code puts the OSR in x and subtracts 1 from it, jumping directly to send. Regardless of the value of x, the send part labels the next instruction anyway.

The send part is just pushing this all back and storing the new value back in the OSR:

```
send:
    mov isr, x
    push noblock
    mov osr, x
    jmp read
    """
```

The final instruction loops back to read to recheck the sensor. There is a wrap directive in other PIO dialects that would save an instruction here; however, CircuitPython PIOASM does not implement this at the time of writing.

That was a lot of assembler language. This example is in the chapter repo as pio_encoder_counting.py.

We need a little more Python code to assemble the instructions, run it, and get the data:

```python
assembled = adafruit_pioasm.assemble(program)

left_enc = rp2pio.StateMachine(
    assembled,
    frequency=0,
    first_in_pin=board.GP20,
    jmp_pin=board.GP21,
    in_pin_count=2
)

right_enc = rp2pio.StateMachine(
    assembled,
    frequency=0,
    first_in_pin=board.GP26,
    jmp_pin=board.GP27,
    in_pin_count=2
)
```

Here, we create two state machines from the same code! We pass C1 to the state machine as the first input pin, and C2 to the state machine as `jmp_pin`, for each encoder based on *Figure 6.4*. We have also set `frequency=0`, which tells the state machine to go at full speed.

We can still use one buffer to read the two state machines alternately. However, this buffer needs to be type `i` (lowercase) to use signed numbers and count negatively. We also will make two variables to hold the left and right wheel states:

```python
buffer = array.array('i', [0])
left_data = 0
right_data = 0
```

In our main loop, we can start with the left sensor, check whether there is data waiting in the RX FIFO queue with `.in_waiting`, and print both sides if there is:

```python
while True:
    if left_enc.in_waiting:
        left_enc.readinto(buffer)
        left_data = buffer[0]
        print(left_data, right_data)
```

Note that there is no special format now; we are just printing the raw number of counts. The right side is the same:

```
if right_enc.in_waiting:
    right_enc.readinto(buffer)
    right_data = buffer[0]
    print(left_data, right_data)
```

If you run this, you should be able to turn either wheel and see output like this:

```
36    77
36    78
36    79
36    80
36    81
36    82
```

We can now count pulses for both wheels. So, if you make a complete wheel revolution, you should land close to plus or minus 8344 – proving our previous calculation.

You will see that one wheel makes the count go backward. Each motor effectively turns in an opposite direction from the encoder's perspective. We will account for this later.

### Troubleshooting

This example is a lot of code and could go wrong in various ways:

- If the code counts up/down randomly and not consistently, you may need to change `jmp_pin` to the other input pin.
- It could also mean you have missed putting `x` in `isr` before sending.
- Check against the source code from git.

You should now be up and running and getting counts.

## Making reusable encoder code

Because we will reuse this, we will put it into a module and pick it up in our `robot.py`. We will then use this to make a demonstration program.

Let's take what we made previously and put it into a module named `pio_encoder.py`. The following code should all be familiar:

```
import rp2pio
import adafruit_pioasm
import array

program = """
    set y, 0
    mov osr, y
read:
    mov x, y
    in null, 32
    in pins, 2
    mov y, isr
    jmp x!=y, different
    jmp read
different:
    in x, 31
    in null, 31
    mov x, isr
    jmp !x, c1_old_zero
c1_old_not_zero:
    jmp pin, count_up
    jmp count_down
c1_old_zero:
    jmp pin, count_down
    ; fall through
count_up:
    mov x, ~ osr
    jmp x--, fake
fake:
    mov x, ~ x
    jmp send
count_down:
    mov x, osr
    jmp x--, send
```

```
send:
    mov isr, x
    push noblock
    mov osr, x
    jmp read
"""


assembled = adafruit_pioasm.assemble(program)
```

We need a way to create the state machines with their parameters and a wrapper for getting the data. A Python class is an excellent way to do this:

```
class QuadratureEncoder:
  def __init__(self, first_pin, second_pin, reversed=False):
    """Encoder with 2 pins. Must use sequential pins on the
board"""
    self.sm = rp2pio.StateMachine(
        assembled,
        frequency=0,
        first_in_pin=first_pin,
        jmp_pin=second_pin,
        in_pin_count=2
    )
    self.reversed = reversed
    self._buffer = array.array('i', [0])
```

We will call it QuadratureEncoder, as it should work with those types regardless of the mechanism. Inside the class is an __init__ function, which tells Python how to make an encoder object – it takes two pins as its parameters and uses them to create the state machine. The object also makes a buffer to store the most recent return value. Note that the two pins must be in sequence.

There's also a reversed parameter; this is so we can account for one motor turning the opposite way. We cannot just swap pins in the code here, as the in instruction requires pins in sequence.

Next, we need a method to read from the encoder or the old value if there's no change:

```
  def read(self):
    while self.sm.in_waiting:
      self.sm.readinto(self._buffer)
    if self.reversed:
```

```
            return -self._buffer[0]
    else:
        return self._buffer[0]
```

By checking the `in_waiting` state, this reading will not block and only update the buffer if there's a new reading; this is a `while` loop because we only want the most recent FIFO data. It returns the element in the buffer, returning the negative version if the motor is reversed.

We can now add these encoders to the `robot.py` library from the end of *Chapter 5, Driving Motors with Raspberry Pi Pico*. Let's add encoders to the imports:

```
import board
import pwmio
import pio_encoder
```

The new code is highlighted. We can also set up the two encoders. Add the bold code after the motors:

```
right_motor = motor_A1, motor_A2
left_motor = motor_B1, motor_B2

right_encoder = pio_encoder.QuadratureEncoder(board.GP20,
board.GP21, reversed=True)
left_encoder = pio_encoder.QuadratureEncoder(board.GP26, board.
GP27)
```

When we use our robot, we can now use `robot.left_encoder.read()` and an equivalent command for the right encoder to get an encoder reading, which we will now use in a demonstration.

## Measure counts for a known time

We will turn this into a demonstration to see what the count is when driving for a second. Because we have put work into preparing our `robot.py`, this code is simple. Put the following code in `measure_fixed_time.py`:

```
import time
import robot

robot.set_left(0.8)
robot.set_right(0.8)
time.sleep(1)
robot.stop()
print(robot.left_encoder.read(), robot.right_encoder.read())
```

This code loads the `time` library and the `robot` helper. It drives forward at 0.8 speed for 1 second. It then stops and prints readings from each encoder. While this code sleeps, the encoder code is still running.

Making the encoders this simple to use means we can integrate them with more complex behaviors later; this is a good strategy for most sensors.

To run this code, be sure to send the `pio_encoders.py` library, the updated `robot.py`, and then `measure_fixed_time.py`. Remember to update `code.py` to load it, and you should see the following:

```
code.py output:
4443 4522
```

You have begun to take sensor readings from your robot, learning PIO on the way!

## Summary

In this chapter, you have learned about measuring distance traveled using encoders, including the different types of encoders.

You saw the output that quadrature encoders create and how to interpret this as a turning direction.

You were introduced to the powerful PIO state machines present within Pico and saw how you can give tasks such as handling encoders to them.

You brought this together to create a reusable handler for the encoders, and we had a demonstration to see them working.

In the next chapter, we will plan and buy more devices for our robot, leading to more sensing, and remotely drive it.

## Exercises

These exercises can improve your robot and let you practice your skills:

- You have been able to get readings and a count for each wheel when driving for a fixed time. How could you make code that stops the motors after a fixed number of counts? You may need to check the encoder readings in a loop regularly.

- You may have noticed imbalances in the counts – this is normal and due to motor and wheel differences. One way you could improve this would be to design and make a holder for the breadboard with the styrene rod so that it doesn't slide around on the platform.

- Could you write code to slow a motor if it's overtaken another one in its count?

# Further reading

These further reading items will help you continue your studies:

- Raspberry Pi has the definitive reference in *Chapter 3* of their datasheet on using PIO and its architecture: https://datasheets.raspberrypi.com/rp2040/rp2040-datasheet.pdf. They also have PIO code examples in *Chapter 3* of their C SDK document (including MicroPython but not CircuitPython samples): https://datasheets.raspberrypi.com/pico/raspberry-pi-pico-c-sdk.pdf.

- Adafruit documentation for rp2pio is at https://circuitpython.readthedocs.io/en/latest/shared-bindings/rp2pio/ and is worth consulting for its use, along with their *Introduction to CircuitPython RP2040 PIO* at https://learn.adafruit.com/intro-to-rp2040-pio-with-circuitpython.

- A video by YouTuber StackSmasher has a great deep dive into PIO, its architecture, and programming at https://youtu.be/yYnQYF_Xa8g.

# Planning and Shopping for More Devices

**7**

We now have a beginner robot platform up and running. The robot can drive around, and we have a pair of sensors already wired in. However, it becomes more interesting and useful if we add other sensors and devices – perhaps also a way to control the robot remotely!

In this chapter, we will look at some of the devices we will use and what types they are, learning more about robot sensors in general. Then, we will look at actual device part numbers and plan where to put our devices on the robot, where there is space for them, and test-fitting them. Next, we will look at a purchase list to build this. Finally, we will build a sensor bracket for our robot.

In this chapter, we're going to cover the following main topics:

- Introducing sensors
- Choosing device types
- Planning what to add and where
- Shopping list – parts and where to find them
- Preparing the robot

## Technical requirements

This chapter requires the following software and computer setup:

- A computer with the internet
- A printer
- FreeCAD

The following robot-related hardware is required:

- The robot build from previous chapters and a Micro USB cable
- A sharp or fresh-blade plastic cutter
- A metal ruler
- Pin vise drill with 0.5 mm, 2 mm, and 3 mm **High-Speed Steel (HSS)**/twist bits
- Safety goggles
- A flat work area with good lighting, free of interruptions or being nudged
- Standoff or mounting kit with M2 and M3 standoffs, bolts, and nuts
- Screwdrivers with appropriate ends for the bolts
- M2 and M3 spanners to tighten bolts and standoffs

You can find all the FreeCAD designs, along with printable templates for this chapter, on GitHub at `https://github.com/PacktPublishing/Robotics-at-Home-with-Raspberry-Pi-Pico/tree/main/ch-07`.

# Introducing sensors

Sensors are how our robot collects information. You've already seen and used one – the encoders. You were also introduced to absolute versus relative sensors in *Chapter 6, Measuring Movement with Encoders on Raspberry Pi Pico*, so what additional sensors can we consider? And how do we interface with them?

Sensors collect information from devices on the robot, making closed-control feedback loops. Sensors can also collect information about the world around the robot, what is present there, or how it has changed in response to the robot's motions.

## Analog sensor types

We briefly talked about analog and digital in *Chapter 1*. Analog sensors create a varying voltage, whereas digital sensors output only 1s and 0s – binary – using two fixed voltages.

Raspberry Pi Pico has a 12-bit **Analog-to-Digital Converter (ADC)** supporting analog sensors connected to 4 pins. Analog inputs are suitable for simple light sensors and variable resistors as inputs. However, analog sensors are less repeatable than digital sensors, and usually, one pin equals one sensor.

## Timed pulses

Another way that sensors can vary continuously is by using timing. For example, the HC-SR04 ultrasonic distance sensor outputs a pulse based on the detected distance. Timing this pulse provides the reading. These tend to need dedicated pins. These pulses can be measured using the PIO system we saw in the last chapter. These are digital in the sense that they only use binary inputs.

## Data bus sensors

Groups of digital pins can form buses, addressing and reading from many devices, including sensors. They have a complexity cost but allow for more interesting data. These have controllers of their own, often containing calibration data to account for variations in sensor manufacturing. In addition, the controllers do some sensor decoding and processing, reducing the code you'll need. Example buses are USB, I2C, Serial, and SPI.

## The robot block diagram

We can use a block diagram for our robot to get a simplified view of our robot and what we will connect to it.

It may look like the wiring schematic, but it shows the logical relationships between things instead of physical connections or placements. The following diagram shows where the robot is now before we add more sensors to it:

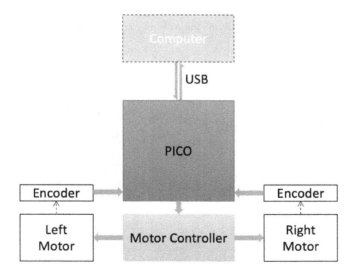

Figure 7.1 – Block diagram of the robot

The preceding figure shows our robot in block diagram form. In the middle, in a darker shade, is Raspberry Pi Pico, the central controller for this robot. Pico connects to the motor controller, which drives the motors. The motors have dashed lines to the encoders since they indirectly influence them, and the encoders, in turn, send data back to Pico, forming a feedback loop. Finally, with a dashed outline at the top of the diagram is your computer, connected via USB. It is dashed here because this is a temporary connection – notice there is a bidirectional arrow here.

We will add more blocks to this diagram as we enhance our robot. Let us see the devices we will be adding.

# Choosing device types

We will add other sensors and Bluetooth to communicate with our robot. We will look in depth at each in later chapters, but we should know enough to consider which we will buy. This section will involve taking a brief overview, making some trade-offs, and choosing the parts to use.

## Distance sensors

Distance sensors, briefly mentioned in *Chapter 1*, let the robot sense its situation and surroundings to avoid or follow obstacles. Some kinds only return if a distance has crossed a threshold, but the more suitable types return a sensed distance value. We will focus on this latter type.

Most distance sensors bounce a beam from objects and measure their return time or angle to determine the distance. For example, clap opposite a wall in a large open space, and you will hear how long your echo takes to return. If you move further from the wall, the return time will be longer.

These fall into two major categories: sound-based and light-based. Each has pros and cons. Let's see some common types:

Figure 7.2 – Distance sensor types

The first sensor in the preceding figure is the VL53L1X, a light sensor. These sensors are lighter and significantly faster than the sound ones and use invisible infrared light. However, some can be confused by other sources of IR light, including bright sunlight.

The right-hand sensor is a sound-based distance sensor, the HC-SR04, identifiable by the two cans. It uses ultrasonic sounds outside the normal range of human hearing. Sound sensors can be confused by secondary sounds, vibrations, and some surfaces – for example, fabrics.

The following table compares their attributes:

| Sensor Type | HC-SR04+ Ultrasonic | VL53L1X IR Light |
| --- | --- | --- |
| Module size | 45 mm x 20 mm x 12 mm | 19 mm x 19 mm x 3.2 mm |
| Weight | 8-10 grams | 3-5 grams |
| Pin usage | 2 per device, not sharable, specific | 2 I2C pins, sharable with other I2C devices |
| Speed | Slow | Fast |
| Max range | 4 m | 1.2 m |

Table 7.1 – Distance sensor comparison

Based on the attributes in the table, the light-based sensors are more suitable for our Raspberry Pi Pico robot due to being smaller and lighter and having the ability to share an I2C bus. We will use two of these sensors facing forward on either side to determine where the closest object is and to avoid it.

Distance sensors give the robot an awareness of its surroundings, but what about the robot's orientation? We'll see a sensor for this in the next section.

## Inertial measurement unit

An **Inertial Measurement Unit (IMU)** lets us detect the robot's orientation. It can give us an absolute orientation and usually combine three sensors:

- A gyroscope – measuring relative rotations
- An accelerometer – measuring accelerations, and using gravity, figuring out which way is down
- A magnetometer – measuring magnetic fields, looking for magnetic poles

Code can combine the data from these sensors to create an orientation relative to gravity and the magnetic north. Some also include a temperature sensor. We describe the IMU sensor count with **Degrees of Freedom (DOF)**. Each combination of direction and type of measurement is a DOF – for example, acceleration on the $x$-axis is one, and rotation on the $x$-axis is another. Having all three sensors is described as 9-DOF, so we should stick with these types.

IMU devices usually use the I2C bus and can share it with other devices, including the distance sensor shown previously.

Effective use of an IMU requires **sensor fusion**, combining calibrated readings from multiple sensors into a helpful form – in this case, an absolute orientation of the robot. Some sensor modules have this ability on board. You want this to offload the handling from Raspberry Pi Pico, and you will not need to write the code for this fusion.

The BNO055 is a great choice; it is common, and has libraries for CircuitPython and some convenient interfaces, with an onboard processor doing the sensor fusion for you.

We now have considered a few sensors, but what about control? The next device we will look at is Bluetooth.

## Bluetooth devices

We have been driving our robot tethered to a computer for starting programs and getting feedback; this is not ideal. However, we can add Bluetooth capability to our Raspberry Pi Pico and communicate to it from our computer or even a smartphone wirelessly.

In *Chapter 9, Teleoperating the Raspberry Pi Pico Robot with Bluetooth LE*, we will consider other options and trade-offs we can make, discussing alternative solutions to Bluetooth **Low Energy** (**LE**). While it is not the only option, the AdaFruit Bluefruit LE UART Friend ADA 2479 module achieves a balance between complexity and cost that works for this project.

This module uses two pins to communicate, with optional pins for extended functionality. We will use two pins for a serial port, also known as a **Universal Asynchronous Receiver-Transmitter** (**UART**). This allows us to send and receive data easily.

It will be used in *Chapter 9, Teleoperating the Raspberry Pi Pico Robot with Bluetooth*, for remote driving and sensor monitoring with Bluetooth.

## Device pin usage summary

We require 2 I2C buses for all these devices – at 4 pins each, for the IMU and distance sensors, and 1 UART with 2 pins for Bluetooth, this consumes a total of 10 pins. Our existing motors used 4 pins each, totaling 8 pins. We have used 18 pins from Raspberry Pi Pico's potential 30, which means you have plenty of pins for expanding this robot.

Since we have ideas for remote operation and sensors, we can consider adding them to our robot.

# Planning what to add and where

The product pages for these devices usually include their dimensions. Depending on which exact breakout you buy, you may need to adapt these designs. Let us use some known models and make a rough test fit to turn into a design.

## Bluetooth and IMU mounting plan

The Bluetooth and IMU should be above the rest of the robot. The IMU's magnetic sensors should not be near the motors and encoders. Putting the Bluetooth higher improves the signal. This rough drawing shows the plan:

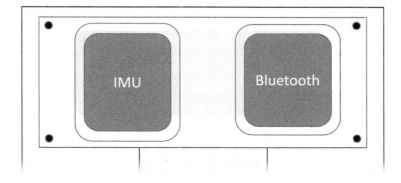

Figure 7.3 – Rough sketch of the shelf

The Bluetooth and IMU, shown in the figure as darker boxes, could be placed on a shelf to distance them from interference with the motors. This shelf, shown in transparent white in the preceding figure, is mounted on standoffs, shown as darker bolt holes. This rough part is superimposed on a FreeCAD sketch to show where it goes.

We are likely, at least while we are developing this robot, to change the wiring more frequently than the batteries, so we can mount the shelf above the batteries using the standoff kit we've already bought.

Next, we need to plan the distance sensors.

## Distance sensor mounting plan

The distance sensors should be at the front of the robot. The following rough sketch shows how they could be mounted:

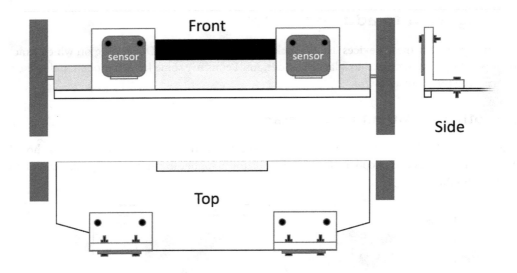

Figure 7.4 – Rough sketch of the distance sensors mounted on the chassis

We can use brackets to mount two sensors facing forward, as shown in *Figure 7.4*. We can make these by sawing sections from right-angled cover trim for walls. We will use bolts to hold the sensor on the bracket and the bracket on the robot. We can add slots for the pin headers.

The headers for the distance sensors are male, and we have a breadboard with female holes. To connect these, we will need more than the precut wires. Male-to-female jumper wires should do for this.

The shelf and brackets are a little more complicated than the styrene rod construction before. We will look at how we put them together below – but first, let us shop.

## Shopping list – parts and where to find them

We've collected enough information to buy the parts we'll need. Let us see what exact parts I recommend.

You bought parts in *Chapter 1*, and used some of these items in previous chapters. However, you should still have stock of the following:

- Styrene 3 mm sheet – these usually come in packs of a few, so you'll have a few around
- Standoff kits with 2 mm, 2.5mm, and 3 mm standoffs, bolts, and screws
- 2.54 mm pitch straight breakaway single-row headers

I then recommend the following additional parts:

- The Adafruit 2742 BNO055 IMU
- The Adafruit Bluefruit LE UART Friend ADA 2479 module
- 2 x Pimoroni PIM373 VL53L1X **Time-of-Flight** (**ToF**) sensor breakouts
- Male-to-female extension jumper wires
- 30 mm PVC right-angle cover trim

We are also going to need some additional tools as we make more interesting parts:

- A medium hacksaw or tenon saw to cut the cover trim.
- A clamp to hold the plastic when sawing.
- A set of needle files for making wiring slots.
- Optionally, some side cutters may make it easier to make the slots.
- Digital calipers for making finer measurements in small spaces.
- About 200 mm strip of wood, roughly 30 mm wide by 20 mm tall to assist in cutting the trim. An offcut will do, but it must be straight.

You can use the same suppliers from *Chapter 1*, *Planning a Robot with Raspberry Pi Pico*, for the parts and tools. Now that we know what to buy, we need to plan how to mount it in more detail.

## Preparing the robot

We have parts and rough ideas for how to mount our sensors.

Using the techniques learned in *Chapter 3*, *Designing a Robot Chassis in FreeCAD*, we can model these brackets and shelves in FreeCAD. You can also get these designs from GitHub at `https://github.com/PacktPublishing/Robotics-at-Home-with-Raspberry-Pi-Pico/tree/main/ch-07`.

The following image shows what this looks like in FreeCAD 3D View:

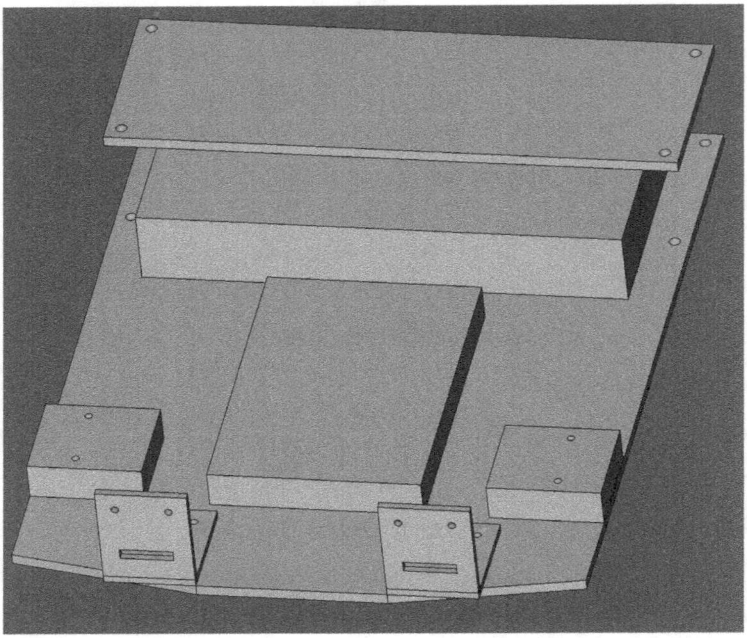

Figure 7.5 – The chassis with the sensor mounts

In the preceding figure, there is a 3D view of the robot in FreeCAD. At the rear, above the batteries, is a shelf for the Bluetooth and IMU. There are bolt holes under this in the chassis.

At the front of the figure are the two brackets for the distance sensors, with mounting holes for the sensors and slots cut out for the connection header to poke through.

Let us take a closer look at each sensor mounting design.

## Designing the shelf

We will make the shelf with styrene. The following diagram shows the suggested dimensions for the shelf layer:

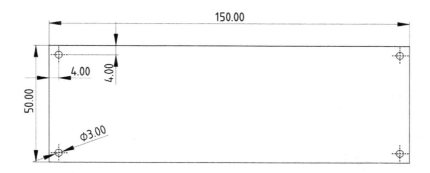

Figure 7.6 – Rear shelf drawing

This diagram shows the shelf. The width of the lower chassis drives the top 150 mm dimension. The side dimension should be enough for the parts to rest on, but you could base it on the scrap styrene material you have left from *Chapter 4, Building a Robot around Pico*, if that is slightly smaller. Just ensure that the bolt holes match the shelf and the chassis. We can model them as a single part.

You can sketch this part in **Sketch Main** on the **XY** plane. Use **Chassis Outline** as a guide with external geometry. Make sure the supports will not interfere with the battery box. The pockets are defined by the dimensions of the IMU and Bluetooth modules, as found on their product pages, with a few extra millimeters for wiggle room.

This shelf needs to be 30 mm above the chassis. The following figure shows how:

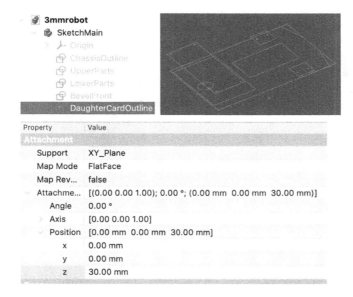

Figure 7.7 – The shelf sketch

Select the sketch for the shelf outline (**DaughterCardOutline**). *Figure 7.7* shows the sketch selected in the **Model Data** tab. The top right shows what we are aiming for with sketch positioning. In the **Property** tab, expand **Attachment**, then **Position**. Set **z** to 30.00 mm. This value will place the shelf above the chassis and batteries.

This sketch can be padded into 3D using the same ShapeBinder and pad techniques discussed before. The following figure shows how this looks:

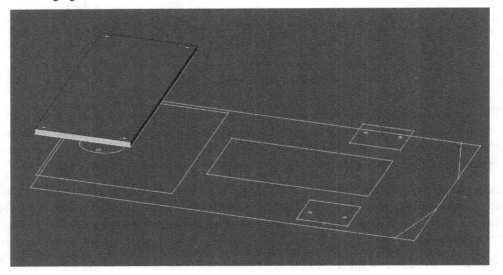

Figure 7.8 – 3D padded shelf

This image shows the shelf padded to 3 mm in 3D.

Don't forget to use ShapeBinders and a pocket to cut the holes in the ChassisPlate part.

Let's see how to build the shelf part from these designs.

## Cutting the shelf

We can use the techniques from *Chapter 4, Building a Robot around Pico*, to cut this. Let's start by creating a FreeCAD drawing to use as a template. The following figure shows the drawing and how to use it to make the part:

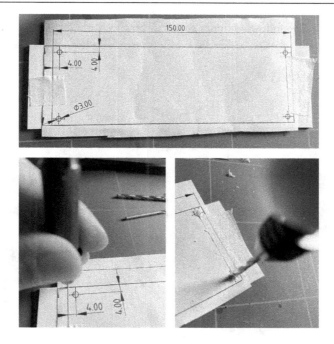

Figure 7.9 – Drilling the shelf

The left part of the preceding image shows the drawings taped carefully onto the part. If the tape is slipping, try using a glue stick instead. Ensure that you've secured the paper template to the plastic and that you can remove it again later.

As we did in *Chapter 4, Building a Robot around Pico*, use a tiny drill bit to dot the corners of these rectangles. We drill the outline dots through the paper and then follow up with the drill holes (3 mm) using a hand drill, leaving the paper in place. Placing some old wood underneath saves you from leaving holes in the cutting mat.

The holes give us a guide. The following figure shows us how we should cut it:

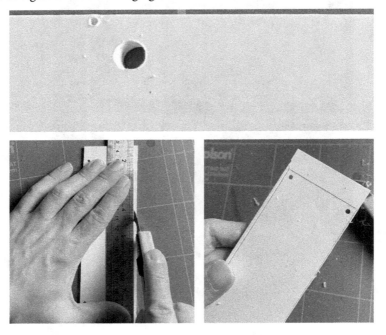

Figure 7.10 – Cutting the shelf

The top of *Figure 7.10* shows the dot for cutting alongside a larger 3 mm hole. We have removed the paper, as some of the lines will be very close to the edge when using scrap. These cuts will be easier without the paper.

Next, as the bottom-left panel shows, line up the metal ruler with two of the dots for an edge – for example, the top line of the left inside pocket – and using a fresh, sharp blade, let the point of the blade find the hole in the top-left corner. You should then be able to follow the ruler and line nearly – but not all the way – to the other hole. Again, this cut is a light score, and we'll make many cuts coming from both sides to keep it between the corners.

Cut until the scores go nearly all the way through. The following figure shows how to extract it from the sheet and finish this part:

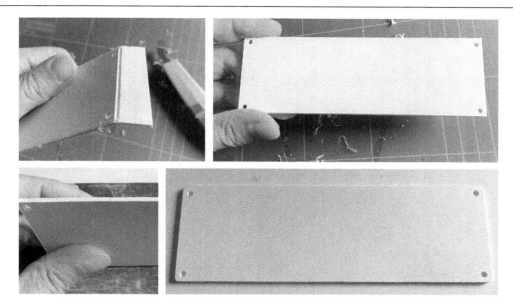

Figure 7.11 – Finishing the shelf

*Figure 7.11* shows the stages of finishing the part. In the top-left panel, flexing the panel has snapped off a section – when scored deeply, this will snap quite nicely. The closer edges may require you to cut through with the knife.

The top-right panel shows this part snapped out of its sheet fully. It is a bit sharp and has some burrs, so we sand it as shown in the bottom-right panel. Low grit (600) should be enough to take off most of the burrs and round the corners for a nicer finish.

You should end up with something like the bottom-right panel. This shelf there looks ready to use.

Now that you have the rear shelf ready to assemble, we can look at designing the front sensor brackets.

## Designing the front sensor brackets

Using our rough designs and the dimensions of the right-angle trim, we can turn these plans into more formal designs and drawings. The following figure shows a close-up of the sensor brackets with dimensions:

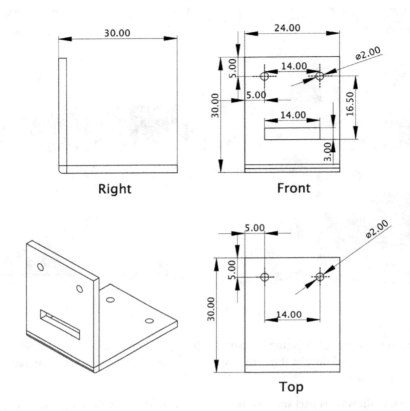

Figure 7.12 – Drawings of the sensor brackets

The figure shows the side, front, and top views of the bracket. There are M2 bolt holes on both sides and a rectangular slot on the front to let through the header connections. The height and length of the brackets depend on the right-angle trim size. The bracket is 24 mm long to accommodate the Pimoroni distance sensor. I used calipers to measure the holes in the Pimoroni sensor for this drawing.

I suggest sketching the top view on the chassis plate, with one rectangle plus a construction mirror line between the holes – that way, you can make shape binders for the base of the bracket. I've aligned the front of these with the front line of the chassis and padded the bracket bases. Next, you make pockets for the holes in `ChassisPlate` and the bracket.

Sketch these brackets symmetrically, as you did for the motors, and make them 42 mm apart. Then create a sketch on the **XZ** plane for the front of the brackets (100 mm forward). Next, pad the tall part up 30 mm from the bases and use the XZ sketch to cut pockets. The following picture shows the brackets in 2D on the chassis plate and how this looks in 3D:

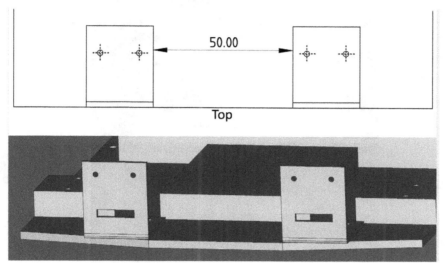

Figure 7.13 – The brackets design on the chassis

The top part of the figure shows a drawing of the brackets dimensioned and lined up on ChassisPlate. The lower panel shows a 3D view of these parts on the model.

This design is enough to create the drawing shown in *Figure 7.13*, and in the next section, we will use it to make this part.

## Cutting the sensor brackets

We can now take this design to the workshop and cut these parts.

The following figure shows how to mark the first two cuts:

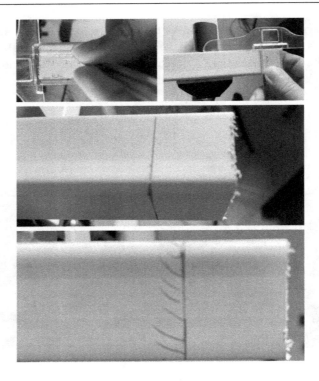

Figure 7.14 – Marking out bracket cuts

The figure starts with using a ruler to mark 24 mm for the width of the bracket. I am using the clamp to keep the angle trim in place. Then, follow up with the square to get a straight line. The middle picture shows this line going around the part.

The last image shows the waste side shaded, so you know where to cut. A cut is not 0 mm, so err on the side of it being too big. Look at the following figure for information on cutting:

Figure 7.15 – Cutting the sensor brackets

The top panel of the preceding figure shows clamping the right-angled trim with the strip of wood supporting it. This wood should stop short of the line you cut at and supports the angled trim while you cut it.

The bottom panel shows me cutting in on the right-angled bend, which is easier than cutting by the sides.

The material will flex as you get to the end and have cut through one side. You need to support the other end and slow down the saw strokes when this happens.

Cut 2 of these 24-mm lengths. These parts will be rough and require finishing with sandpaper, as shown in the following figure:

Figure 7.16 – The parts ready for smoothing

As the preceding figure shows, the parts have very rough edges. You will need to use multiple passes of the grades of sandpaper to smooth both the brackets as the figure shows. For a nice finish, you could round the corners.

We will need to drill out holes as shown in the following figure:

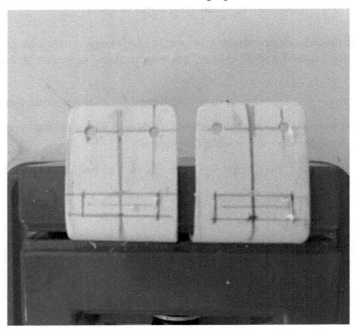

Figure 7.17 – Drilling holes into the sensor brackets

*Figure 7.17* shows the parts in a clamp and marked with a pattern to assist drilling – this has a middle line, then holes on either side. I use calipers for this; however, my drawings are not perfect, and yours do not need to be either.

Start each hole with a smaller drill bit, then open it to 2 mm. This material, uPVC, is tougher than styrene and takes more effort to get through. There are drill holes on the front and through the base of this bracket. Make these before continuing to the slots.

The toughest part will be drilling out the slots. The following figure shows how to cut this slot out:

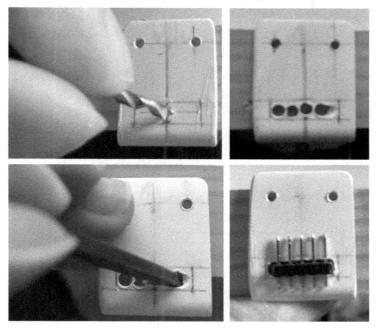

Figure 7.18 – Cutting and filing out the slots

The top-left panel in the preceding figure shows me making a hole with a 3-mm bit. Make holes along the length of the slot, as shown in the top-right panel. They do not need to be perfectly constrained in the slot but closely inside.

As the bottom-left panel shows, choose a needle file with a corner (angular) profile, and push it into the holes to break them out. You could optionally use side cutters to join the holes. Then, when you have joined the holes, file the slot out into a rectangle. Finally, when you have made a wider shape, use a wide flat needle file to make a letterbox profile.

The bottom right shows that the hole only needs to be big enough to accommodate the plastic pin header body. At that point, you can stop filing.

We've now manufactured the shelf plates and the brackets. However, we still need to modify the chassis plate and fit them.

## Preparing the chassis plate

The chassis plate needs eight additional holes to accommodate the new fittings. You should be able to use your FreeCAD designs to get the following drawing:

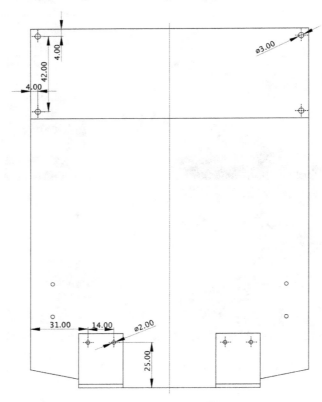

Figure 7.19 – Drawing of the chassis plate with sensor mounting holes

At the top of this drawing, representing the robot's rear, is the shelf, with 4 additional 3 mm holes to drill, 2 on either side, for putting in the shelf standoffs.

At the bottom of the figure (the front of the robot) are four additional holes at the front of the chassis, 2 mm wide for bolting on the brackets.

For this part, we have marked out the dimensions on the drawing, and we will use calipers to mark these holes before drilling them, as shown in the following figure:

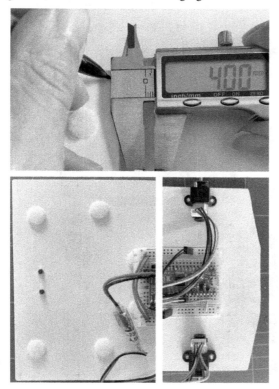

Figure 7.20 – Measuring and marking out the chassis plate holes

In the top panel of *Figure 7.20*, I use the straight edge at the top of the calipers to mark one of the rear shelf holes at 4 mm from the edge. The panel in the bottom left shows the holes for the rear shelf marked out this way, and the bottom right shows the holes for the brackets.

While not essential, marking and drilling the chassis plate with the upper deck parts and wheels removed is easier. When you have completed these steps, just be sure to put them back as specified in earlier chapters.

You can drill out the holes in the usual way, starting with the smaller diameter drill bit, then making the bracket holes 2 mm, and the shelf holes 3 mm. After drilling the uPVC, you will notice how much easier they are to drill!

We are now ready to assemble the robot again.

## Assembling the robot

We can now get into the exciting business of adding new sections, starting with the shelf, as shown in the following figure:

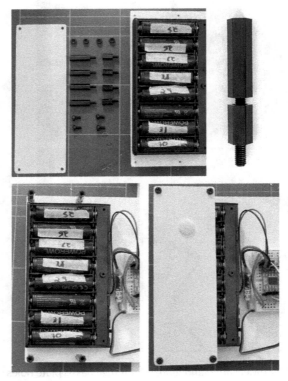

Figure 7.21 – Assembling the Bluetooth and IMU shelf

The top-left panel in the preceding figure shows the robot alongside the shelf panel and a set of standoffs. As shown in the top-right panel, you may need to join two standoffs to clear the battery box.

As the bottom panels show, you then bolt the standoffs into the robot base, with nuts to support them underneath. Bolt the shelf layer on top of the standoffs; this should look like the bottom-right panel. I've also added a hook-and-loop dot for the Bluetooth module. We will leave the IMU side to add standoffs.

Now, we just need to add in the sensor front-facing brackets, as the following figure shows:

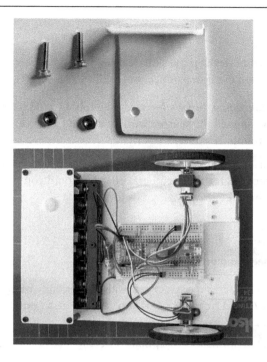

Figure 7.22 – Assembling the front-facing brackets

As the top of *Figure 7.22* shows, we need an M2 bolt and nut for each bracket. Bolt these into the holes at the front of the chassis.

The chassis now looks like the bottom panel. This robot is now ready for us to start experimenting with sensors. We will not bolt the sensors in just yet, as we have wiring to do for them, which we will get to in the following chapters.

## Summary

In this chapter, we learned about more sensor and device types. We then revisited our robot, planning where to add these sensors.

The chapter showed a shopping list for these parts and then used dimensions from the product pages of these parts to make CAD drawings. The CAD drawings let us visualize what we wanted in 3D. Next, we fabricated parts from these drawings.

We revisited manufacturing parts and used a variation on the template technique to make the brackets.

Finally, we assembled all these parts and now have a robot base ready to add sensors. We will wire in and program these sensors in the following chapters, starting with the distance sensors.

## Exercises

These exercises will help further develop the skills learned in this chapter. As we have not attached the switch to the chassis, we can use it to practice these skills:

- The sensor brackets are facing directly forward. The robot would have better sensor coverage and look better if you aligned the sensors with an angled front. This will likely require using other FreeCAD constraints or skills.

- The shelf is flat and uses a hook-and-loop to hold devices. Using FreeCAD, could you design another layer with recesses for the peripheral boards? Could you then use the skills learned here to build it? You may require a coping saw for this.

## Further reading

We referenced these datasheets and product pages in the text:

- Adafruit Bluefruit UART LE Friend product page with dimensions – `https://www.adafruit.com/product/2479` and `https://learn.adafruit.com/introducing-the-adafruit-bluefruit-le-uart-friend/downloads`

- Adafruit BNO055 breakout page with dimensions – `https://www.adafruit.com/product/2472`

- Pimoroni VL53L1 breakout page with dimensions – `https://shop.pimoroni.com/products/vl53l1x-breakout`

For further information on Bluetooth LE, consider *Building Bluetooth Low Energy Systems* by Muhammad Usama bin Aftab. This dives into the details of wireless network communication systems suitable for use in the **Internet of Things** (**IoT**). IoT concepts translate well into robotics.

To find out more about sawing, check out the following guide:

- *A Little Saw – A Workshopshed Guide to Cutting Tools* by *Andy Clarke* has excellent information on cutting different materials well. This little book shows the suitable saws and the right ways to use them.

# 8
# Sensing Distances to Detect Objects with Pico

Our robot is starting to move around independently. We spent the last chapter preparing mount points to add sensors, including distance sensors. We can use these sensors to detect how far objects are from the robot, and by adding more than one, we can see which direction is closest. This sense will allow the robot to respond to the real world and drive around a room without much manual control.

In this chapter, we will learn more about these sensors and their limitations. Then, we will attach the sensors to the robot and learn more about the communication protocol used to talk to them. Next, we will wire the sensors into Raspberry Pi Pico and get data. Finally, we will tie multiple sensors together with motor control to make the robot avoid obstacles while driving.

In this chapter, we're going to cover the following main topics:

- How distance sensing works
- Soldering headers and attaching them to the robot
- An introduction to I2C communication
- Communicating with a single distance sensor
- Connecting two distance sensors
- Building a wall avoider with Raspberry Pi Pico

# Technical requirements

This chapter requires the following:

- The robot build from previous chapters
- 2 x Pimoroni VL53L1X distance sensor modules
- 2 x five-way single row 2.54-mm (included with the modules)
- 8 x male-to-female jump wires with a 2.54-mm DuPont connector
- 2 x M2 nuts and M2 x 6-mm bolts
- A suitable screwdriver for driving the bolts, and a spanner for holding the nuts
- A soldering station with a soldering iron, solder, tip-cleaning brass, and soldering stand
- A flat work area with good lighting, free of interruptions or being nudged
- The code from previous chapters
- A Raspberry Pi Pico code editor such as Mu or Thonny
- A USB micro cable

You can find the code for this chapter at `https://github.com/PacktPublishing/Robotics-at-Home-with-Raspberry-Pi-Pico/tree/main/ch-08`.

# How distance sensing works

Before diving into connecting and programming distance sensors, we should examine how they operate. *Chapter 7, Planning and Shopping for More Devices*, evaluated options and chose optical (light-based) distance sensors. We will be focusing on this type for the remainder of this chapter.

Many distance sensors operate using a principle known as **time of flight**. The following diagram demonstrates this:

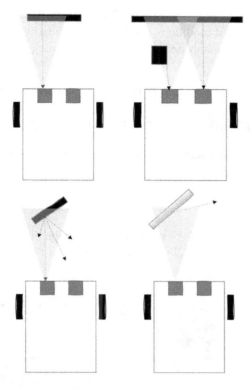

Figure 8.1 – Optical distance sensor operation

The preceding diagram shows pictures of robots with sensors and the returned light. On the top left, a single beam is emitted (shown as a cone), hits an object, and its reflection (shown as a dashed line) hits the sensor (the blue box), which detects it. The time between emitting the beam and receiving the response is the time of flight used to calculate the distance. At the top right of the diagram, both sensors are active. However, the left sensor detects a closer object in its beam, returning a lower value than the right sensor, which only detects the wall behind the object.

For most materials, the light beams make a diffuse reflection in all directions, as the bottom-left part of the diagram shows, and these time-of-flight sensors will detect their return. However, like the mirror shown in the bottom right of the diagram, some materials do not diffuse their reflected light and reflect away at the incident angle. Therefore, these objects may not be detected or will confuse the system. It may even detect objects reflected in a mirror as behind the mirror.

Note that these sensors have a slight instability in their readings, so they will take several readings and combine them to produce a more accurate reading. The sensors use a timing budget to take several readings. The VL53L1X device does this sampling and averaging.

Ambient light conditions can affect these sensors, with very bright light potentially washing out the beam. However, these work in most conditions, losing accuracy and a little distance but not becoming unusable. In our context, a robot mainly needs to detect an oncoming obstacle, so a loss of accuracy is acceptable.

We now understand how the sensors operate. In the next section, let's attach them to the robot.

# Soldering headers and attaching them to the robot

Before we can start to use or wire the sensors, we will need to solder headers onto them and then bolt them so they face forward on the robot.

## Soldering headers

I recommend using a spare breadboard for soldering these, as you did with Raspberry Pi Pico and the motor controller earlier in *Chapter 4, Building a Robot around Pico*. The following photo shows me soldering them:

Figure 8.2 – Soldering the distance sensor headers

The photo on the left of the preceding figure shows the sensors, the headers, and a breadboard to aid soldering. Place the headers long pins in the breadboard holes and the sensors on top. Pimoroni designed these sensor modules to hold the board on the header for easy soldering.

In the right photo, one sensor has had the header soldered in, and I am soldering the other. The headers should be facing back from the sensor so that the wiring will not be in the sensor beam.

With the headers soldered, you are ready to fit the sensors onto the robot.

## Mounting the sensors

We put a lot of groundwork into mounting the sensors in previous chapters and created brackets to mount them onto. The following photo shows these sensors bolted on:

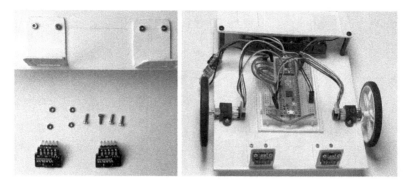

Figure 8.3 – Optical distance sensors mounted

On the left of the preceding photo are the robot's front, the sensors, and the M2 nuts and bolts. We push the sensor's headers through the slot and bolt them in place. Note that there might be a slight tightness around the slot. If so, file the header space up and out a little to accommodate this. Wipe or blow away any dust after filing. The right shows the sensors bolted onto the front of the robot.

We will wire these sensors in via I2C. But, first, let's take a closer look at how I2C is used to talk to sensors like this.

# Introduction to I2C communication

You encountered I2C communication in earlier chapters. *Chapter 1*, discussed how I2C is a data bus that carries address information, allowing a primary device such as Raspberry Pi Pico to reach multiple devices on a single bus. We learned then that Raspberry Pi Pico has two hardware I2C buses. **I2C** (or **I²C**) is an acronym for **Inter-Integrated Circuit**.

In *Chapter 7, Planning and Shopping for More Devices*, we saw how we would be using I2C devices both for VL53L1X distance sensors along with an IMU.

How exactly does this bus work? *Chapter 1*, also mentioned that I2C has two wires – a **Serial Clock** line (**SCL**) and a **Serial Data** line (**SDA**). The following picture shows how devices send signals through them:

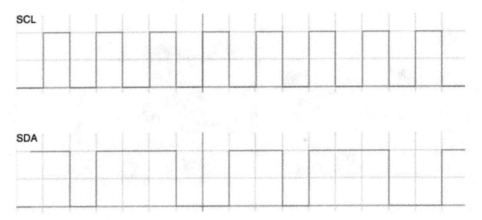

Figure 8.4 – I2C signals on the wire

The preceding diagram shows two graphs representing I2C signals. The horizontal axis is time, and the vertical axis when high is logic one, with low being logic zero. As shown in the top diagram, the clock produces a stream of pulses, a square wave. The accompanying data line sends data synchronized with the clock pulses. This clock means that devices on the bus are synchronized.

The lines are usually held high (logic high) and pulled down when a device wishes to communicate. Devices leave the line high when they have stopped communicating so another can control the bus. In most cases, the central controller will send a request to a device, and the device will respond.

As mentioned in *Chapter 1*, devices on an I2C bus have an address. However, these VL53L1X devices both have the same addresses. Luckily, there are two I2C buses we can use on Raspberry Pi Pico.

The good thing is that we don't need to control much of this manually. CircuitPython has a `busio` (bus input-output) library for handling I/O operations on a data bus, which we'll use to control these devices.

We need to wire both SDA and SCL for each sensor and then write code connecting them. In the next section, we will discuss how to wire and talk to a single sensor.

## Communicating with a single distance sensor

Each distance sensor requires only four wires; however, we will also improve the power system. We will then get into the code needed to read data from a system.

## Wiring the distance sensors

We start our wiring by looking at a block diagram of our robot, as we saw previously in *Chapter 7*. Look at the following diagram:

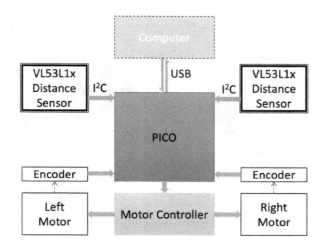

Figure 8.5 – The robot block diagram with distance sensors

The preceding diagram shows the robot block diagram with the additional VL53L1x distance sensors connected via I2C to Raspberry Pi Pico. The new parts have a thick double outline.

We need the schematic to get into the details of the connections, as shown in the following figure:

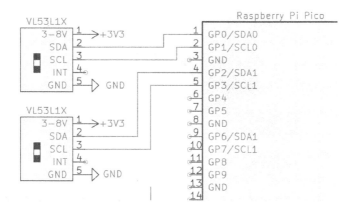

Figure 8.6 – The schematic with distance sensors

In the preceding figure, I've shown a close-up schematic of the distance sensors connected to Raspberry Pi Pico. We connect a sensor to each I2C bus, along with 3v3 power and ground connections.

The right sensor has its SDA connected to GPIO0 and its SCL connected to GPIO1. The left sensor connects SDA to GPIO2 and SCL to GPIO3.

The following photo shows the sensor wiring:

Figure 8.7 – Distance sensors wired into the robot

The preceding photo shows the robot with the sensors wired in using male-to-female jumper wires. Remember that you are not connecting the INT pin from the sensor to anything, so expect a gap here. Also, double-check power and ground connections, as reversing these may damage the device. It is also a common troubleshooting problem with I2C to have accidentally swapped SDA and SCL lines.

Once the wiring is complete, please carefully remove the protective cover from the sensor, as shown next:

Figure 8.8 – Removing the protective tape

As the preceding photo shows, if your sensors still have protective cover tape, carefully use a fingernail or tweezers to remove this cover before use. After removing this, take care not to touch the optical parts of the sensor.

With this wired in, let's look at how the sensor operates in the next section.

## VL53LX theory of operation

These sensors have a particular operation cycle. We can use this as a guide for writing our code. The following diagram shows the states of the VL53L1X:

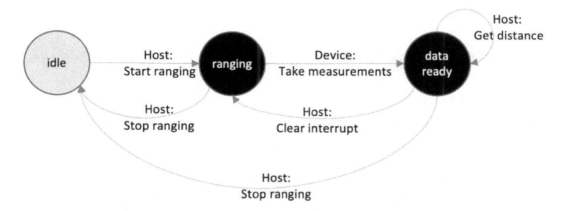

Figure 8.9 – VL53L1X operating states

The preceding state diagram is a simplified view of the states of this device. The nodes are the states, and the lines between them are events/signals that cause the state to change. First, the sensor starts on the left in an idle mode. This mode saves power until we want to measure distances. Then, the host (Pico) sends a start-ranging signal, and the device enters ranging mode, where it takes active measurements. Next, the device enters a data-ready state when it has taken enough measurements, as set by the timing budget. In this state, the host can read the distance measured.

However, the device will not make fresh measurements until the host sends a clear interrupt signal, putting it back in the ranging state.

Finally, when we no longer need ranging, the host should send the stop-ranging signal, which puts the device back into an idle state and uses less power.

We will need to account for these state changes in our code, which we will cover in the next section.

## Reading a single distance sensor in CircuitPython

When reading a sensor, we will use the Adafruit VL53L1X library. Copy `adafruit_vl53l1x.mpy` from the Adafruit CircuitPython library into the `lib` folder on the `CIRCUITPY` volume. We can also write code inspired by their documentation examples. When communicating with any new device, using the example code for the related library is always a good start. We will adapt it a little for Raspberry Pi Pico.

We start the code in the read_1_sensor.py file with imports:

```
import time
import board
import busio
import adafruit_vl53l1x
```

We've two new imports here. Alongside the vl53l1x library, the busio library uses Pico pins to form data buses such as I2C.

Next, we need to set up the I2C bus and device:

```
i2c = busio.I2C(sda=board.GP2, scl=board.GP3)
vl53 = adafruit_vl53l1x.VL53L1X(i2c)
```

This first line forms the I2C bus using GP0 and GP1 pins, corresponding with the left sensor. Pico is a little fussy about which pins you can use for the SDA and SCL lines of an I2C bus. We then create the VL53L1X device with this I2C bus. In the next section of code, we send some settings to the device:

```
vl53.distance_mode = 1
vl53.timing_budget = 100
```

The device has multiple distance modes; distance mode 1 is short-range – for close objects. We set a timing budget of 100 milliseconds, controlling how long the sensor is allowed to take for measuring. It limits the maximum distance and the number of measurements used to smooth data.

Let's start the device:

```
vl53.start_ranging()
```

We have now moved the device from the idle state into the ranging state. We can now wait for the measurements to be ready. We can now write the code for the main loop:

```
while True:
    if vl53.data_ready:
        print("Distance: {} cm".format(vl53.distance))
        vl53.clear_interrupt()
    time.sleep(0.05)
```

This loop starts by checking whether the device has data ready to read. When it has data_ready, we can read data with vl53.distance and print it. This distance is in centimeters.

After reading the data, we need to send a clear_interrupt signal, so the sensor goes back into its ranging mode for a new reading.

Regardless of whether there is data, the system will wait 50 milliseconds before looping around and checking again.

Upload this file to Pico and update `code.py` to import `read_1_sensor.py`. Then, when you run it and connect the REPL, you should see output like this:

```
code.py output:
Distance: 35.1 cm
Distance: 34.4 cm
Distance: 33.8 cm
Distance: 35.2 cm
```

Let's check a few things to ensure everything's all working before we carry on.

## Troubleshooting

If this example code doesn't work, the following troubleshooting steps will help:

- If you receive the `No pull up found on SDA or SCL; check your wiring` warning, this indicates the wiring may be incorrect or that wires may be loose. Power down and check the wiring.

- The `No I2C device at address: 29` warning likely means you have reversed SDA and SCL. Swap them and try again.

- There must be no heat in any part of the circuit including the wires, the batteries, the sensor, or Pico – this will cause damage, and should be powered down then the wiring checked carefully.

- Sometimes, sensors can get stuck between test runs, resulting in slow or erratic results or sensors showing no errors but never having data ready. I advise powering the whole robot down.

- Try adding lines such `print("i2c set up")` between stages to see where the problem is.

- If you see `unsupported operation` or `unknown distance mode`, check the power wiring to ensure each sensor is well connected.

Note that powering down means turning off battery power (if you've turned that on) and unplugging the computer. Never leave the computer plugged in during rewiring.

We should have data being read reliably from a single sensor. Our robot has two sensors, though, so let's read data from both in the next section.

## Connecting two distance sensors

We have wired in two sensors, each on a separate set of pins. Create the `read_2_sensors.py` file. The imports look identical:

```
import time
import board
import busio
import adafruit_vl53l1x
```

When we come to set up the sensors, we first need to set up two I2C buses on the different pins and then use them:

```
i2c0 = busio.I2C(sda=board.GP0, scl=board.GP1)
i2c1 = busio.I2C(sda=board.GP2, scl=board.GP3)
vl53_l = adafruit_vl53l1x.VL53L1X(i2c0)
vl53_r = adafruit_vl53l1x.VL53L1X(i2c1)
```

We can also apply the same configuration settings for both sensors:

```
vl53_l.distance_mode = 1
vl53_l.timing_budget = 100
vl53_r.distance_mode = 1
vl53_r.timing_budget = 100
```

The main loop starts in the same way, with both sensors going into ranging mode:

```
vl53_l.start_ranging()
vl53_r.start_ranging()

while True:
```

And in this case, we will check for ready data from both sensors before printing:

```
    if vl53_l.data_ready and vl53_r.data_ready:
        print("Left: {} cm, Right: {} cm".format(vl53_l.
distance, vl53_r.distance))
        vl53_l.clear_interrupt()
        vl53_r.clear_interrupt()
    time.sleep(0.05)
```

When uploaded and run on Pico, this code will now output both sensors' data, as shown next:

```
Left: 39.2 cm, Right: 37.4 cm
Left: 38.8 cm, Right: 37.4 cm
Left: 39.0 cm, Right: 37.4 cm
Left: 38.8 cm, Right: 37.6 cm
Left: 39.1 cm, Right: 37.6 cm
```

We'll troubleshoot any problems here before we use two sensors for smarter behavior.

## Troubleshooting

Adding a second sensor can still cause odd behavior:

- First, the wiring considerations in the single sensor apply.
- If you have used the same I2C bus twice in code, you will get some odd behavior or the same reading from both sensors. Check the code for this.

We now have data from two sensors, and in the next section, we'll use this data to make the robot avoid objects.

# Building a wall avoider with Raspberry Pi Pico

Two distance sensors and independent motor control with some code are the ingredients needed to avoid obstacles. Let's start by putting the distance sensors in the shared robot library.

## Preparing the robot library

Like we have with other aspects of the robot, we'll start by building the distance sensors set up in the robot.py file. At the top of this file, the imports now include busio and adafruit_vl53l1x libraries:

```
import board
import pwmio
import pio_encoder
import busio
import adafruit_vl53l1x
```

We can then set up our left and right distance sensors. Insert the following below the encoder setup and above the `stop` function:

```
i2c0 = busio.I2C(sda=board.GP0, scl=board.GP1)
i2c1 = busio.I2C(sda=board.GP2, scl=board.GP3)

left_distance = adafruit_vl53l1x.VL53L1X(i2c0)
right_distance = adafruit_vl53l1x.VL53L1X(i2c1)
```

Save this file and be sure to upload it to the CIRCUITPY volume.

We will use `robot.py` in the avoider code. We must next consider how avoiding behaviors operate.

## Wall-avoiding theory of operation

We need to consider data from both sensors to avoid walls. The following diagram shows how we will do this:

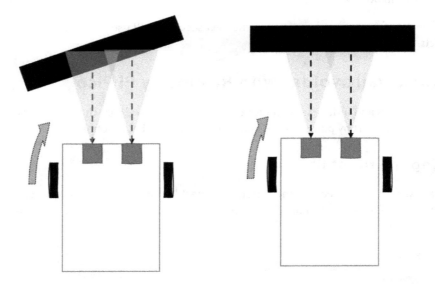

Figure 8.10 – Avoiding walls with two sensors

In the preceding diagram, a sketched robot is facing a wall. It has sensed the wall, but the object is closer to the left sensor than the right; dashed arrows show reflections coming from the object. That means the robot should turn right to avoid this object, shown by the curved arrow to the robot's left. We will make the turn by reversing the right motor until it is clear of the obstacle.

There is a special edge case where the two sensors detect the wall as equally close. To ensure the robot decides to avoid the flat obstacle, we will slightly bias (weight one side) by checking one sensor first and making a turn based on that.

We have all the parts we need. So, let's start writing this code.

## Distance sensor wall avoider code

The code for this uses the distance sensors from the robot library. Put this in `avoid_walls.py`. Let's start with familiar imports and by setting the sensor config:

```
import robot
import time

robot.left_distance.distance_mode = 1
robot.right_distance.distance_mode = 1
```

We'll leave sensors on the default timing budget. We then have some configurations for our avoider:

```
too_close_cm = 30
speed = 0.9
```

The `too_close_cm` variable has a threshold for when the robot should turn to avoid a wall. We can set the overall robot speed for this behavior in the `speed` variable. We can tune these two variables to ensure the robot avoids a wall in time. Let's start the sensors ranging:

```
robot.left_distance.start_ranging()
robot.right_distance.start_ranging()
```

We are going to start the robot moving; however, we want to ensure that the robot stops moving and stops the sensors ranging if there are any problems, so we wrap the main loop in `try`:

```
try:
    robot.set_left(speed)
    robot.set_right(speed)

    while True:
```

A `finally` statement to accompany that `try` will come below the main loop. Next, we check whether there is sensor data ready:

```
    if robot.left_distance.data_ready and robot.right_distance.
data_ready:
        left_dist = robot.left_distance.distance
        right_dist = robot.right_distance.distance
```

We store the read distances so that we can use them throughout the handling. Note that while `robot.left_distance.distance` looks like a variable, it is a property that actively reads the sensor when we use it.

Since we have two values, we should check if one side is too close. Note that a timeout will result in a 0 value, so we should check values are above this too. By favoring a side, we slightly bias the robot to that side, and this should stop the robot from being *indecisive* if both sensors detect a close obstacle:

```
    if 0 < right_dist < too_close_cm:
        print("Obstacle detected - Left: {} cm, Right: {} cm".
format(left_dist, right_dist))
        robot.set_left(-speed)
```

We check whether the distance on the right sensor is closer than the threshold. If so, we print a line of debug, showing that we've detected an obstacle and the two sensor readings. We then set the left motor to go backward, which will cause the robot to swerve left, away from the obstacle.

We can now handle what happens otherwise:

```
    else:
        robot.set_left(speed)
        if 0 < left_dist < too_close_cm:
            print("Obstacle detected - Left: {} cm, Right: {}
cm".format(left_dist, right_dist))
            robot.set_right(-speed)
        else:
            robot.set_right(speed)
```

If we've not turned left, we ensure the left motor is going forward. We then check the left distance sensor, and if this is too close, we turn right. Finally, we set the right motor forward, so both motors will be going forward if it detects nothing too close.

Putting the left distance check in the else means that the robot will not set both motors backward and will favor turning left in front of an obstacle directly in front.

We then need to finish the loop by clearing the interrupts (so that the sensors are ranging again), and we leave a little time for them to sense again:

```
robot.left_distance.clear_interrupt()
robot.right_distance.clear_interrupt()
time.sleep(0.1)
```

Now, we need to handle any errors that happened. If you recall, we wrapped this code in a try block. The finally block stops and cleans everything up:

```
finally:
  robot.stop()
  robot.left_distance.clear_interrupt()
  robot.right_distance.clear_interrupt()
  robot.left_distance.stop_ranging()
  robot.right_distance.stop_ranging()
```

The first thing we do here is to stop the motors. If anything goes wrong, such as errors from sensors or the code, we don't want the robot to drive into a wall. We then clear the sensor interrupts and stop the sensors, putting them back into idle mode.

This example is complete, and you can send it to the robot along with the updated robot.py. Keep the battery power turned off initially and test that it detects obstacles. You should see output like the following:

```
Obstacle detected - Left: 29.6 cm, Right: 42.2 cm
Obstacle detected - Left: 17.2 cm, Right: 39.6 cm
Obstacle detected - Left: 10.4 cm, Right: 39.1 cm
Obstacle detected - Left: 5.2 cm, Right: 42.6 cm
Obstacle detected - Left: 6.7 cm, Right: 42.3 cm
Obstacle detected - Left: 18.4 cm, Right: 27.6 cm
```

Once you've seen it working and detecting, you can turn the battery power on, which will enable the motors and let drive along – either using a long USB cable or independently if it's working well. Let's check whether we have any problems with some troubleshooting.

## Troubleshooting

These steps should get you up and running if you have any problems:

- If the robot complains about importing modules, ensure you have uploaded the code and libraries for this chapter and previous ones.

- If the robot is getting too close before turning, try increasing the `too_close` variable.

- With any behavior using motors, ensure the batteries are fresh. Low power can cause stalling motors and malfunctioning sensors.

- The sensors will not detect obstacles that are above or below them. This limitation means the robot will drive into and get stuck on low obstacles or under high ones.

Your robot now avoids walls and objects.

## Summary

In this chapter, we learned about distance sensors. We looked at how the sensors operate and how to attach them.

We learned more about the `I2C` bus and then saw how to electrically connect these VL53L1X distance sensors.

We then looked at the operating modes of the VL53L1X sensor and wrote code to get readings from one. Finally, we finished with a behavior to avoid walls using this sensor.

In the next chapter, we will gain remote control of our robot by adding Bluetooth.

## Exercises

You can use these exercises to practice more of the concepts learned in this chapter:

- Could you write code to follow an object at a fixed distance? If it's further away, could you drive it forward, and if it's too close, back it up a little?

- Try different materials in front of the robot, such as glass, black fabric, thin paper, and thick paper. Observe which are detected and how they affect the distance detected. For example, what happens if you aim the robot at a mirror?

- Try observing the distance measurements in different light conditions, such as room light, darkness, and full sunlight.

# Additional reading

The following sources of additional reading can deepen your understanding of these sensors:

- `https://github.com/adafruit/Adafruit_CircuitPython_VL53L1X` has the library and sample code for driving a device.

- `https://www.ti.com/lit/an/sbau305b/sbau305b.pdf` is a datasheet for a different brand of optical sensors. However, it has excellent information on reflectance and how materials affect distance sensing.

- `https://www.st.com/en/imaging-and-photonics-solutions/vl53l1x.html#documentation` contains the complete product documentation. Of particular interest are the product specifications and API user manual.

For an alternative technique using ultrasonic distance sensors, refer to *Learn Robotics Programming – Second Edition* by *Danny Staple*.

# 9

# Teleoperating a Raspberry Pi Pico Robot with Bluetooth LE

We intend for the robot we are building to be mobile. We already have the robot driving on the floor and able to sense and respond to its surroundings. However, we either rely on it blindly or are tethered to it with a laptop. Neither is quite what we want. What if we could get feedback while it's untethered and roaming the floor?

In this chapter, we'll see how Bluetooth **Low Energy** (**LE**) is well suited to this task, allowing us to get data from the robot, use an app to graph data, and even remotely control our robot from our smartphone!

In this chapter, we will cover the following main topics:

- Wireless robot connection options
- Connecting Bluetooth LE to Raspberry Pi Pico
- Making a Bluetooth LE sensor feed on Raspberry Pi Pico
- Teleoperating the robot with Bluetooth LE

## Technical requirements

This chapter requires the following:

- The robot from *Chapter 8, Sensing Distances to Detect Objects with Pico*
- An Adafruit Bluefruit LE UART Friend ADA2479
- 1 x eight-way single-row 2.54-mm header (included with the module)
- 5 x male-to-female jump wires with a 2.54-mm DuPont connector
- Access to an Android or iOS smartphone
- Velcro hook and loop dots

- The code from previous chapters

- A Raspberry Pi Pico code editor such as Mu or Thonny

- A USB micro cable

You can find the code for this chapter at `https://github.com/PacktPublishing/Robotics-at-Home-with-Raspberry-Pi-Pico/tree/main/ch-09`.

## Wireless robot connection options

So far, we've been working with the robot tethered to our computer. We send code to it and use the REPL tools to see what it is doing or printing out. While the REPL tools can be convenient, having a wire between the computer and the robot is not so convenient and limits how far the robot can drive or has you running behind it with the laptop. The following diagram shows how we could do things:

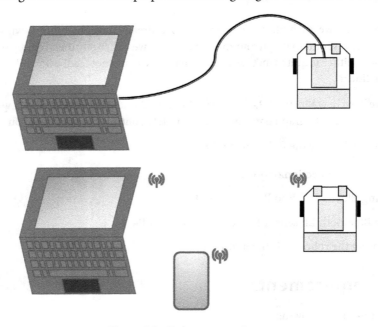

Figure 9.1 – Robot connections

The top part of the diagram shows things tethered with a wire. But the bottom part shows that the computer and the robot are not physically wired together. Instead, they are using wireless to send data to each other.

Once we are wireless, we can also consider a smartphone coming in as an alternative item. We can use a wireless medium to send data from the robot's sensors or code to see what is going on and monitor it. We can also send control signals to take control and drive our robot.

There are several different wireless protocols we can use. All require our robot to have a **transceiver** – transmitter and receiver – board.

While some robot controllers, such as Raspberry Pi 4, have onboard transceivers, Raspberry Pi Pico does not. Therefore, we will need to add a breakout. In addition, they come with different protocols and different implementations of those protocols.

Which transceiver boards might we choose and why? The following table shows the comparison:

| Board | Protocol | Pros | Cons | Cost |
|---|---|---|---|---|
| Adafruit AirLift ESP32 coprocessor | Wi-Fi (and potentially Bluetooth BLE) | • In principle, fast<br>• Flexible<br>• Could serve a web page<br>• Computer/phone support | • Requires local Wi-Fi network<br>• Highest power usage<br>• Highest code complexity<br>• Medium communication latency | $10-20 |
| Carrier module + XBee Zigbee module | Zigbee | • Long-distance, potential mesh network | • Hard to find modules<br>• Code/design complexity depends on the carrier<br>• Medium power usage<br>• Expensive<br>• Unlikely to have PC/phone support | $30-40 |
| Adafruit RFM69HCW transceiver radio | Packet radio/LoRaWAN | • Long-distance | • Low speed<br>• Unlikely to have PC/phone support | $5-10 |
| Bluefruit LE UART Friend | Bluetooth LE (low energy) | • Computer/phone support<br>• Does not require an access point<br>• Prebuilt Adafruit app<br>• Low power<br>• Easy to connect<br>• Low latency<br>• Excellent support | • Less flexible than Wi-Fi<br>• Slow data rate | $10-20 |
| HM-10 BLE | Bluetooth LE (low energy) | • Low<br>• Easy to connect | • Incompatible with other apps – uses unusual protocol implementation<br>• May need custom code to view<br>• Poor support | $10 |

Table 9.1 – Transceiver modules for Pico

In the preceding table, I've picked boards that come with onboard software stacks, reducing the amount of code we need. We are less concerned with speed as we don't intend to send camera data; however, latency is important as we want our robot to respond quickly to commands and send up-to-date sensor data.

Support is important – we want to choose modules with good support from their vendors and the community. For example, Adafruit has excellent support and documentation for their modules, with online communities on Discord and other forums and all their code available on GitHub, which gives their modules a lot of credibility over cheaper and less well-supported options.

An honorable mention must go to Raspberry Pi Pico W – a Pico with an onboard Wi-Fi chip. This has excellent support from the Raspberry Pi community. It has the added complexity of requiring you to serve up a graphing web frontend, however it may make a very good alternative.

The HM-10 modules are widely available and may even be super cheap, but their unusual protocols mean connecting with them needs more code.

The choice with the most going for it here is the Adafruit Bluefruit LE board. It has low current usage and is small. There is a **Serial Peripheral Interface (SPI)** and a **Universal Asynchronous Receiver/Transmitter (UART**; defined in more detail as follows) version of this board. **Bluetooth LE** is a low-energy variant of Bluetooth, ideal for short-range communications between devices such as a robot and a controller. It has a range of up to 100 m. Bluetooth LE has two-thirds of the data rate compared to regular Bluetooth but consumes half the current when active. Smart software profiles allow it to frequently use low-power modes, and rapidly wake up when needed.

UART doesn't need much configuration and only uses two wires (as opposed to the three or more wires SPI or I2C uses). There is no clock line (just an agreement on speeds) and no address, just one-to-one device communication. We have already been using a USB-based UART to communicate with Raspberry Pi Pico.

We will use the Adafruit Bluefruit LE UART Friend board for simplicity in our project.

It is widely available through Adafruit directly or through distributors such as Pimoroni and Mouser Electronics.

Adafruit Bluefruit is an ecosystem of Bluetooth LE-based development boards, so there's lots of compatible code. In addition, it works with computers and phones that have built-in Bluetooth LE transceivers, and Adafruit makes apps for both computers and phones to communicate with them. These apps will save us time, as other solutions require you to build your apps.

We now know which module we will use. So we can make use of it, let's take the Bluefruit LE UART board and connect it to our robot!

## Connecting Bluetooth LE to Raspberry Pi Pico

The Bluefruit LE UART Friend is relatively simple to wire in. First, the module will need headers soldered onto it, and then we can look at how to attach it to the robot physically and how to wire it. We will then connect to it from our robot and a smartphone.

Adding Bluetooth LE will result in our robot having a block diagram as follows:

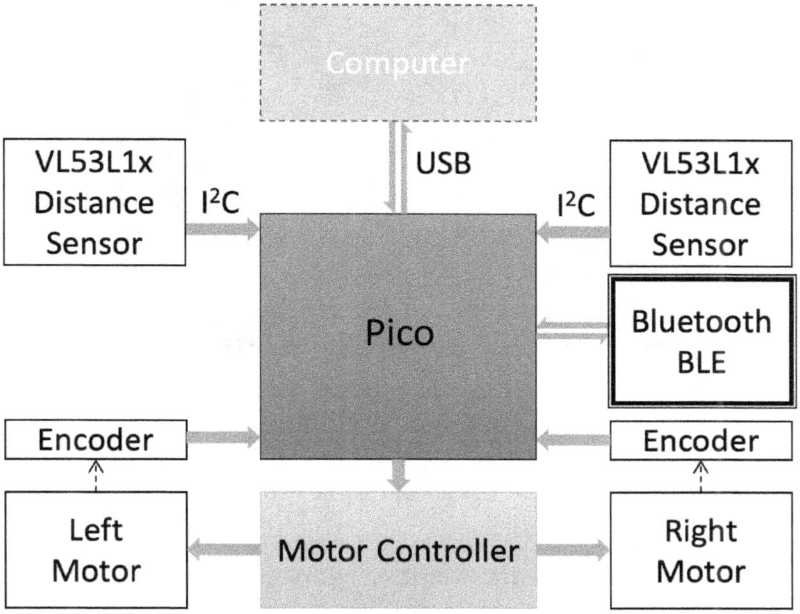

Figure 9.2 – Robot block diagram with Bluetooth

The preceding diagram shows the robot blocks with the additional Adafruit Bluefruit LE UART Friend (marked as a Bluefruit module) connected via UART to Raspberry Pi Pico.

Solder a set of male headers onto the board using the same techniques used for the modules in *Chapter 8, Sensing Distances to Detect Objects with Pico*.

## Attaching the Bluetooth module to the robot

We want to place the module above devices that are most likely to restrict and interfere with the Bluetooth. We created a breakout shelf in *Chapter 7, Planning and Shopping for More Devices*, for this purpose.

The following photo shows how the headers should be attached:

Figure 9.3 – Adafruit Bluetooth LE UART Friend with headers

The previous figure shows the device with the headers. They should be soldered so they are standing above the pin names, facing the same way as the switch. The switch should be in UART mode.

We want the attachment to be good enough not to slide around or stick out awkwardly due to cable tension. You could make a more permanent connection by drilling the appropriate holes in the shelf, but a convenient way for quick prototype platforms is to use Velcro (hook and loop) dots. Look at the following photo:

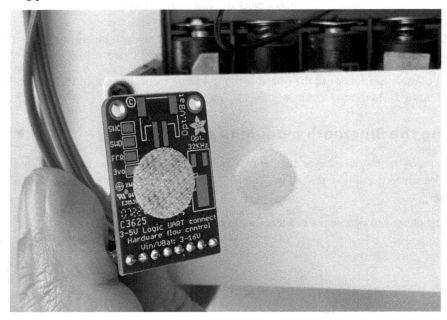

Figure 9.4 – Bluefruit module Velcro connection

The previous photo shows the Bluetooth breakout module with a Velcro dot ready to attach to the Velcro dot already attached to the robot's shelf.

The Velcro gives us a convenient way to attach/detach the module without it sliding off.

Next, we need to wire the Bluetooth breakout into Raspberry Pi Pico.

## Wiring the Bluetooth breakout to Raspberry Pi Pico

Wiring the Adafruit Bluefruit LE UART Friend module requires only five wires. The following photo shows the connections you will need:

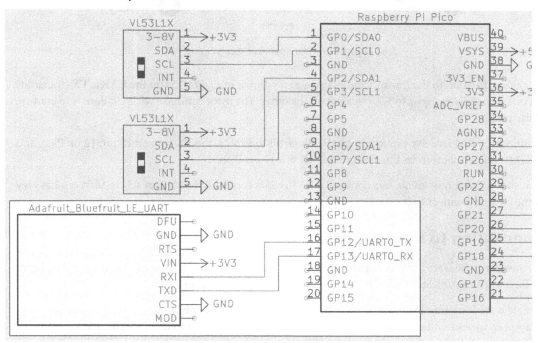

Figure 9.5 – Connecting the Bluefruit LE module to Pico

The preceding diagram shows, as a schematic, the connection between the Bluefruit module and Raspberry Pi Pico.

First, connect the module to power and ground, with the VIN going to 3.3V power. Next, the CTS pin needs to be connected to ground to send and receive data via UART. The next figure shows how the TX and RX pins interact:

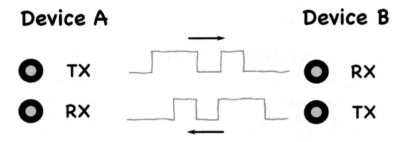

Figure 9.6 – Transmit and receive pins

Pay close attention to the transmit and receive pins (shown in *Figure 9.5*) on the UART. TX (transmit) on one device always goes to RX (receive) on the other. The most common wiring failure is to confuse these pins.

Connect the RXI (receive input) pin on the Bluefruit to PIN12, the TX (transmit) pin 12 on Pico, and TXD (transmit data) from Bluefruit to the RX pin 13 on Pico.

You should have now made five connections. The Bluetooth LE device is ready to turn on. Let's try some code to connect to it over UART.

## Connecting to the Bluefruit LE device with UART

Now we have the board connected and a smartphone ready to talk to it, how can we get our Raspberry Pi Pico to communicate? Let's start by making a Hello Bluetooth app.

Create a folder named `bluetooth-hello-world`. In a slight departure from our previous examples, we can name the main file `code.py`, and we only need to drag and drop the content of our example folders onto the CircuitPy volume.

We start `bluetooth-hello-world/code.py` with imports for the board, and `busio`, which has code for the UART bus:

```
import board
import busio
```

We then use the IO pins from `board` to create a UART object:

```
uart = busio.UART(board.GP12, board.GP13, baudrate=9600)
```

Note that we could have multiple UARTs on different pin combinations for other sensors and devices. We have specified the **baud rate**, the rate at which the signal changes per second. `9600` is the default for this device, as specified in its datasheet.

We can use this in a loop to check for any input – this is how we know something is connected:

```
while True:
    if uart.read(32) is not None:
        uart.write("Hello, Bluetooth World!\n".encode())
```

In every loop, we try reading up to 32 bytes. If we get anything (it's not showing `None`), then we respond with a `hello` message. The `uart.write` method sends bytes, not strings, so we must encode the string to send it. Also, note the `\n` character at the end – this is a new line in the output.

Copy this over to Pico, and it will now be running, waiting for something to connect. So, let's connect something to our Pico via Bluetooth LE!

## Connecting a smartphone

A smartphone makes a great client to connect to the robot and see what is going on. You can use an Android/iOS smartphone to connect to the Bluefruit by finding the Bluefruit LE Connect app on the app store appropriate to that phone.

This app is free. Alternatives are available at `https://learn.adafruit.com/introducing-the-adafruit-bluefruit-le-uart-friend/software-resources`, including desktop apps with similar functionality.

Load the Bluefruit LE Connect app. The following screenshots show what you'll see:

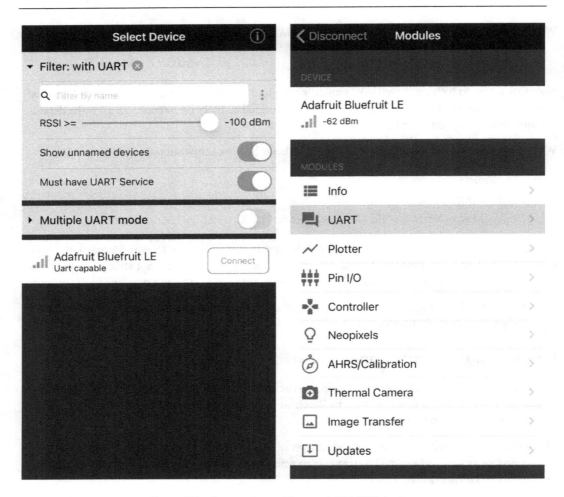

Figure 9.7 – Connecting to Bluetooth LE UART devices

In the preceding screenshots, the panel on the left shows the app. With your robot powered on and the Adafruit Bluefruit LE connected, you should see the device in the list on the app. There may be many Bluetooth devices; you can turn on the **Must have UART Service** toggle to filter these. You can then click the **Connect** button to connect to the device.

When doing so, you should see a solid blue light on the Bluefruit board on the robot. You will also see the screen on the right. Bluefruit LE Connect may ask you to perform an update on the Bluetooth device; if so, please follow the onscreen instructions and accept this before proceeding. This may take a short while.

Click the **UART** button in the menu to send and receive data. You should see screens like those shown in the following screenshot:

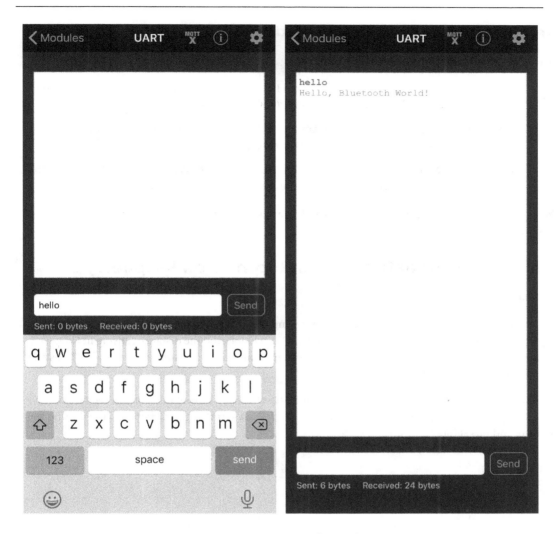

Figure 9.8 – Interacting over Bluetooth UART

The UART screen, shown in the preceding screenshot, lets you interact with and see output from the module. The left panel shows me typing hello. Try this yourself and hit the **Send** button to send data to the module – our code will respond when you send something to it.

The right panel shows the robot responding with the message. It may take around a second to respond here. It may send the message twice if you send more than 32 characters.

You should now be in contact with the module. If not, try the following troubleshooting steps.

### Troubleshooting the Bluefruit module

The following should get you up and running:

- When it is powered up, there should be a red light on the Bluefruit module. If not, disconnect the power immediately and verify the power (3.3V) and ground wiring.

- You should see a solid blue light when you connect to the Bluefruit device from the smartphone. If not, verify that you have connected to the correct device.

- If you cannot see the `Hello, Bluetooth World!` message, please verify that the TX and RX wiring is correct; reversing them is a common issue with the wiring.

You should now have this module connected and able to send data from the robot to a listening device. Let's make use of this to send sensor data.

## Getting sensor data over Bluetooth LE on Raspberry Pi Pico

So far, you've tested the sensor-based examples, seeing their output in the console by connecting your laptop to it. However, building on our `hello world` example and the distance sensing in *Chapter 8, Sensing Distances to Detect Objects with Pico*, we can not only see the sensor output over UART as text but also plot in in a graph. So, let's get into it.

We'll put this code in a folder named `bluetooth-distance-sensors`. Copy in the `robot.py` and `pio_encoder.py` files. We will add `code.py`. Let's start with the imports, combining the sensors and bus setup:

```
import board
import time
import busio
import robot

uart = busio.UART(board.GP12, board.GP13, baudrate=9600)
```

With the UART now prepared, we can prepare the sensors:

```
robot.left_distance.distance_mode = 1
robot.left_distance.start_ranging()
robot.right_distance.distance_mode = 1
robot.right_distance.start_ranging()
```

We've set both sensors ranging and in the correct mode. We can now start a loop and fetch the sensor data:

```
while True:
    if robot.left_distance.data_ready and robot.right_distance.
data_ready:
        sensor1 = robot.left_distance.distance
        sensor2 = robot.right_distance.distance
```

We wait until the data is ready before fetching the distance, and we store the distance from both sensors. We can now send the data to the UART:

```
uart.write(f"{sensor1},{sensor2}\n".encode())
```

We use an f-string to format the data from both sensors into one line, separated by a comma. We must also include the end-of-line, \n, character again and encode it into bytes.

We must then ensure the sensors on the robot are ready to take another reading:

```
robot.left_distance.clear_interrupt()
robot.right_distance.clear_interrupt()
```

We can wait for a little time before trying again. Add the following code outside the if block but inside the while loop:

```
time.sleep(0.05)
```

This sleep completes the code for this example. As an overview, this loop will read the sensors, send data when it has a reading, and then send this over the UART, and it will sleep for a little time and go again.

If you copy this code to the robot and connect the phone with the **UART** menu option, you will be able to see the two numbers vary as you move things in front of the sensors. The following screenshot shows an example:

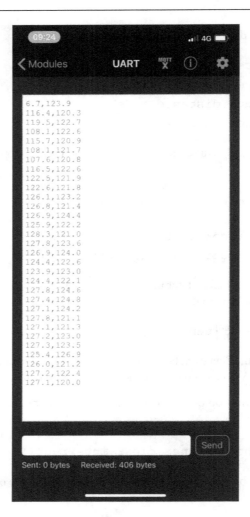

Figure 9.9 – The sensor output as text

The previous screenshot shows how the data is output as plain text numbers. You can disconnect the computer and put the robot on independent battery power, and you should still be able to connect to it and see the sensor readings.

This data feed is great as we have a remote view of the robot. However, we can now go one better and graph this data.

## Graphing the data

The phone app has the built-in ability to graph data in comma-separated format. We can use this for the output of numeric data to quickly visualize what is going on.

From the UART screen, click on the back button to go back to the options menu for this connection. Then, click the **Plotter** button to access the plot mode.

The following screenshots show how to do that:

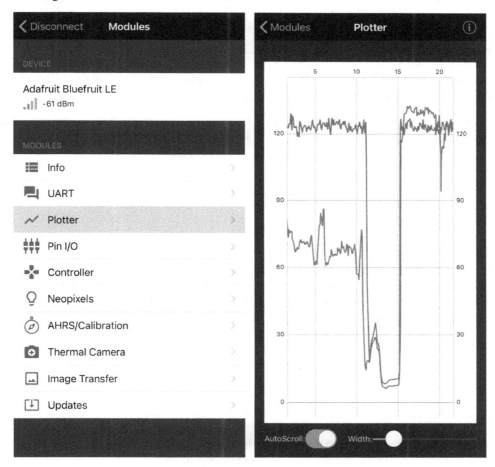

Figure 9.10 – Enabling graphing

The preceding screenshots show how to access the graph functionality and an example sensor data graph.

The app will use any comma-separated numeric data, and I've tested it with six columns of data so far. If the output seems patchy, please ensure you allow the app to make an over-the-air update of the Bluetooth device, as this significantly improves the throughput.

We've seen how to get robot sensor data and use it to plot what is going on with the sensors. However, we might also want to use our Bluetooth services to take control of the robot. Let's see how in the next section.

# Controlling the robot with Bluetooth LE

Bluetooth LE is a two-way medium. In this section, we'll see how to receive data from the UART and, better yet, how to decode that data into control signals. By the end of this section, you'll be able to drive your robot with a smartphone!

## Printing what we got

Before we try to decode control packets, let's just make a simple app to echo whatever shows on the Bluefruit UART out onto the Raspberry Pi Pico console.

Put the following code in `bluetooth-print-incoming/code.py`. We start by importing and setting up the UART port:

```
import board
import busio
uart = busio.UART(board.GP12,board.GP13,baudrate=9600,
timeout=0.01)
print("Waiting for bytes on UART...")
```

The one difference here is that I've added a short timeout. Without the short timeout, the port will wait a full second for the number of bytes read. You might have noticed with the Hello world example that it took a second before you got the output, and this will be why. We want to get control data as soon as possible. There's also a `print` statement, so we know it's ready.

Now we have the main loop:

```
while True:
    if uart.in_waiting:
        print(uart.read(32))
```

This loop reads data from the port. First, it checks whether there is data waiting, then tries to read up to 32 bytes and immediately prints what we got, then tries again.

If you go back to the **UART** menu option in the phone app, you will be able to type messages on the phone and see them appear in the Pico console:

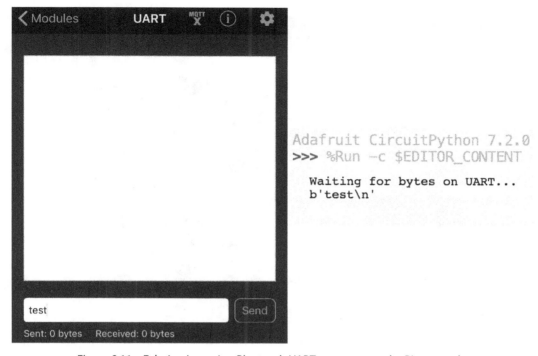

Figure 9.11 – Echoing incoming Bluetooth UART messages on the Pico console

*Figure 9.11* shows a screenshot of the phone on the left, ready to send a message. The screenshot on the right shows the message appearing in the Raspberry Pi Pico console. Notice the b prefix. This prefix means it's a raw byte message. We will be able to extract our button data from this.

In the next section, let's see how the smartphone app can use this for control.

## Button control mode

On the phone app, there were several different menu modes for interacting with the robot. One of these is **Controller** mode. The following screenshots show how this looks:

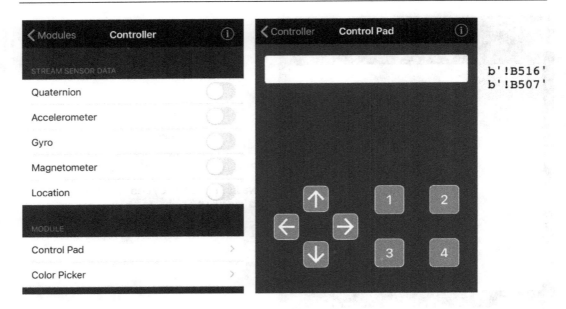

Figure 9.12 – Bluefruit app Controller mode

The screenshots show the controller mode. The leftmost screenshot shows the initial **Controller** app screen; from this, we pick **Control Pad**.

The screenshot in the middle shows the control pad. In this screenshot, on the left is a directional pad, and on the right is a set of numeric buttons. We can send both signals to the robot.

The third, rightmost screenshot shows how the control signals look when printed. This output looks like some strange text, but that is because it is data encoded into a data **packet**. A packet is a chunk of data on a bus.

These control signal packets encode the button that changed and whether it was pressed or released. Try pressing a button now with the `print-incoming` app, and you will see a set of control codes. They aren't particularly human-readable, so we'll need to decode them.

Let's make some code to detect and decode button control packets.

## Decoding button control packets to drive the robot

Adafruit has a library for specifically handling and decoding these packets. Copy the `adafruit_bluefruit_connect` folder from the CircuitPython bundle into your Pico at `CIRCUITY/lib`.

We can import this into our code and use it to drive the robot. Create a folder called `bluetooth-teleoperation` on your computer. Copy the most recent `robot.py` and `pio_encoder.py` files into this. We'll start a `code.py` file with the imports:

```
import board
import time
import busiofrom adafruit_bluefruit_connect.button_packet
import ButtonPacket
import robot
uart = busio.UART(board.GP12,board.GP13,baudrate=9600,
timeout=0.01)
```

The imports are mostly familiar, but we've added the button packet type so we can decode control pad buttons. We also set up the UART to receive data.

The keys send a keypress and key release. What we don't want is for the robot to receive nothing from the phone and keep driving, so we will have a stop time. We are also going to set an overall driving speed:

```
stop_at = 0
speed = 0.8
```

We can now get into the app's main loop. The first thing we'll do is check for waiting data, and if there is some, decode it as button presses:

```
while True:
  if uart.in_waiting:
    packet = ButtonPacket.from_stream(uart)
```

The `from_stream` function will decode a button packet directly from the UART. It frees us from considering the byte size of that packet by trying to read the right number of bytes.

If we have a packet, we can check whether the button was pressed or released and ensure we stop the robot if it is released:

```
      if packet:
        if not packet.pressed:
            robot.stop()
        elif packet.button == ButtonPacket.UP:
          robot.set_left(speed)
          robot.set_right(speed)
        elif packet.button == ButtonPacket.DOWN:
          robot.set_left(-speed)
```

```
        robot.set_right(-speed)
    elif packet.button == ButtonPacket.LEFT:
      robot.set_left(-speed)
      robot.set_right(speed)
    elif packet.button == ButtonPacket.RIGHT:
      robot.set_left(speed)
      robot.set_right(-speed)
```

The preceding code first checks whether you've pressed the button. We then started matching the button code with different buttons and changing motor speeds to drive or turn depending on which you pressed.

Next, we need to consider timeouts. If we have pressed a button, we should reset the timeout, and in the outer loop, we should check the timeout:

```
        stop_at = time.time() + 3


  if time.time() > stop_at:
    robot.stop()
```

At the end of the `if` packet block, we add 3 seconds to the current time; this will be when we time out. Then, at the bottom of the `while` loop, if we've passed the `stop_at` time, we stop the robot.

Copy this over to CircuitPython on Pico, and you will now be able to use the buttons to drive the robot. You can disconnect it from the computer, turn on battery power, and drive it.

You can now use control pad buttons to drive the robot. Press and hold a button and it will drive for up to 3 seconds without a further keypress; you'll need to press multiple times to drive further. This 3-second timeout is a compromise between ensuring it doesn't run away and making it fun to drive.

## Troubleshooting

If the robot is not responding and you have been through the previous examples, try this troubleshooting method.

In code like this, adding `print` statements will help. When you have the robot connected to the computer, you can just use `print`. Otherwise, use `uart.write(message.encode())`. Try adding these before the `while` loop starts and in places where the code handles buttons.

By printing before the `while` loop, we know our code started (if not, we can connect it to the computer to look for error messages). Likewise, we can tell that button signals are being decoded by printing when it handles buttons.

Printing like this lets us narrow down where the problem is. By checking the lines around any output we fail to see, we can see what might be incorrect or whether there is wiring to verify.

You should now have a remote-controllable robot!

## Summary

In this chapter, we have seen how to hook a Bluefruit LE transceiver to our robot and then use it to send and receive data. We've seen the robot data go to a smartphone and data go from a smartphone back to Pico on the robot.

We then took this up a level and sent formatted data to plot sensor information on the phone, allowing us to remotely visualize the robot's state.

Finally, we used the smartphone app to control and drive the robot. In the next chapter, we will look at the PID algorithm, a neat way to tie sensor data and outputs together in a feedback loop, and we'll use our new remote data plotting ability to tune it!

## Exercises

These exercises let you extend the functionality of your robot code and deepen your understanding of the topics:

- In the Bluetooth control app, there are four numeric buttons. Could you extend the control program to use these to control the robot's speed?

- The Bluetooth control pad app also has a little window to show messages. Try sending messages back from the robot code to the app to show in this window.

- Could you use the plotting code with the encoder counts and plot these? Perhaps divide their total counts by elapsed time in the code, or reset the encoder counts and reread them to plot a rate per second.

# Further reading

- Adafruit's great support for the Bluefruit UART LE Friend includes a product page with dimensions – `https://www.adafruit.com/product/2479`. In addition, they have content on the Adafruit learn website at `https://learn.adafruit.com/introducing-the-adafruit-bluefruit-le-uart-friend/`, including material on more ways to connect and use the device, complete datasheets and specifications, along with additional software.

  Adafruit also has a `#help-with-radio` channel on their Discord with a community that specifically helps with problems, questions, and ideas about their transceiver modules.

- For further information on Bluetooth LE, check out *Building Bluetooth Low Energy Systems* by Muhammad Usama bin Aftab. This book has a detailed dive into wireless network communication systems suitable for use in **Internet of Things (IoT)**. IoT concepts translate well into robotics.

# Part 3: Adding More Robotic Behaviors to Raspberry Pi Pico

Now that we've seen some sensors, we can add more interesting robot behaviors. We will learn robot algorithms to make more use of the distance sensors and encoders. Then we introduce the Inertial Measurement unit. Finally, we will look at how to use the sensors to locate a robot in a known space.

This part contains the following chapters:

- *Chapter 10, Using the PID Algorithm to Follow Walls*
- *Chapter 11, Controlling Motion with Encoders on Raspberry Pi Pico*
- *Chapter 12, Detecting Orientation with an IMU on Raspberry Pi Pico*
- *Chapter 13, Determining Position using Monte Carlo Localization*
- *Chapter 14, Continuing Your Journey – Your Next Robot*

# 10

# Using the PID Algorithm to Follow Walls

We built a robot with multiple sensors and used distance sensors in *Chapter 8, Sensing Distances to Detect Objects with Pico*. We can use smarter algorithms with these sensors to make smoother behaviors.

In this chapter, we will investigate the PID algorithm, building its stages into something that will follow objects, then turning that into something that will follow a wall. We will use our new ability to graph over Bluetooth to tune the settings and get a smooth result.

In this chapter, we will cover the following main topics:

- Introducing the PID algorithm
- Using a PID to follow a wall
- PID tuning – using graphs to tune the PID

## Technical requirements

You will need the following items for this chapter:

- The robot and code from *Chapter 9, Teleoperating Raspberry Pi Pico Robot with Bluetooth LE*
- An open space with room to move, and objects suitable for following
- An Android/iOS device with the Bluefruit app
- A hand screwdriver with a 2 mm bit
- Digital calipers
- A pencil

You can find the code for this chapter at `https://github.com/PacktPublishing/Robotics-at-Home-with-Raspberry-Pi-Pico/tree/main/ch-10`.

# Introducing the PID algorithm

In this section, we will introduce the different parts of the PID algorithm while building on what you have already seen.

## Control and feedback

Controlling robot systems generally depends on feedback loops like the following one:

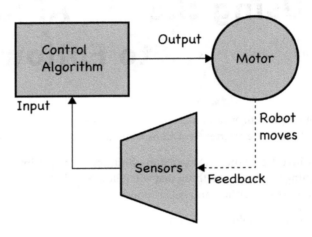

Figure 10.1 – Control and feedback loop

The preceding figure shows data from the sensors going into a control algorithm. The algorithm controls the motor as its output. The motor will cause the robot to move. This movement leads to a feedback loop as the sensor reading changes and goes through the cycle again. This concept is known as **closed-loop control**.

This closed-loop lets the robot interact with the real world, adjusting its behavior to produce the desired result.

We built a simple system like this for our distance sensors. We'll look more closely at that system next.

## Bang-bang control

In the examples provided in *Chapter 8, Sensing Distances to Detect Objects with Pico*, the system compared the distance sensors on our robot against a threshold. For example, look at the following diagram:

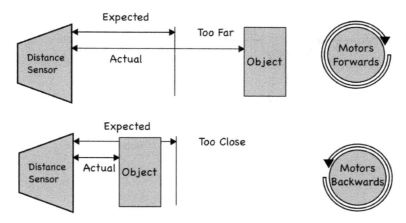

Figure 10.2 – Bang-bang motor control

*Figure 10.2* summarizes **bang-bang control**. This control system has two modes. If the actual measurement is above the expected value and too far away, it will drive forward; otherwise, it assumes that it is below the value and too close and will drive backward. It will always drive in one of the directions and at a fixed motor power. Controlling with only a fixed power is known as **constant correction**.

This method is simple and suitable in some situations, but occasionally, something smoother is needed. What if we wanted the robot's motor power to change depending on how far away it is from the object? We'll need to calculate an error value, as shown in the following diagram:

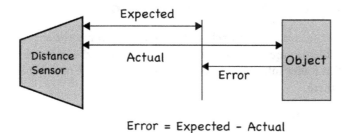

Error = Expected - Actual

Figure 10.3 – Calculating an error value

*Figure 10.3* shows how we calculate the error value. By subtracting the actual value from the expected, we will get the error. This error will change in magnitude, depending on how different things are, and change in direction, depending on which side of the expected measurement the actual measurement falls.

If we multiply this error by some value, this could be turned into a motor speed, such that a higher magnitude error will lead to a larger motor movement. This process will form the feedback control loop shown here:

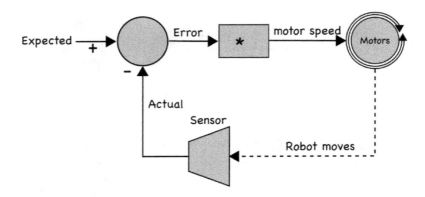

Figure 10.4 – A proportional feedback loop

*Figure 10.4* builds upon *Figure 10.1*, feeding an error value through a multiplier (or gain) to calculate the motor speed, which moves the robot. Robot movement feedback changes the sensor reading. With the right gain value, this robot will slow down as it approaches the object.

Multiplying the error value to control the output like this is known as **proportional control**, which is one part of the system we are building. The amount by which we multiply the error is known as the **proportional gain**.

At this point, I think we are ready to write the code for this.

## Distance sensing with proportional control

In this section, we'll write code to approach an object and maintain the expected distance from the object. Create a folder on the host named `proportional-distance-control`. We'll copy the content of this folder into the top directory on the Pico. We can also copy the robot and `pio_encoder` files there.

Add a `code.py` file, starting with imports and enabling the UART:

```
import time
import board
import busio
import robot
uart = busio.UART(board.GP12, board.GP13, baudrate=9600)
```

Now, we can add our proportional controller as a class:

```
class PController:
    def __init__(self, kp):
```

```
        self.kp = kp

    def calculate(self, error):
        return self.kp * error
```

This code lets us make proportional controller (`PController`) objects with a proportional gain, kp. We can call the `calculate` method with an error value to get the control value.

Let's set up a sensor:

```
robot.right_distance.distance_mode = 1
robot.right_distance.start_ranging()
distance_set_point = 10
distance_controller = PController(-0.1)
```

We make a set point of 10 cm for our expected distance. Then, we have our `distance_controller`, which is using the `PController` object.

When our robot is further away, we need to drive forward, so we need to use a negative proportional gain. The motor speeds are between -1 and 1, so -0.1 will reduce the distance by a tenth and negate it.

The main loop will only check the `PController` when there's a new distance reading:

```
while True:
    if robot.right_distance.data_ready:
        distance = robot.right_distance.distance
```

We can use this with the set point to calculate the error and feed it into the proportional controller:

```
        error = distance_set_point - distance
        speed = distance_controller.calculate(error)
        uart.write(f"{error},{speed}\n".encode())
```

While we're here, we can send the numbers to the UART (so that we can plot them). We have a speed from the error. Now, we can send this speed to both motors:

```
        robot.set_left(speed)
        robot.set_right(speed)
```

Finally, we must reset the sensor for another reading and delay it a little:

```
        robot.right_distance.clear_interrupt()
        time.sleep(0.05)
```

Send this code to the robot. If you place an object in front of the robot, it will find and hold a position about 10 cm from the object. You should also be able to plot the output.

The motors here may be beeping a lot; this is not good for them and will make them hot. We can use code to establish a **dead zone**, where we expect the control output to be too small for the motors to respond and turn them off instead.

Add the following code after the speed calculation to stop the beeping:

```
if abs(speed) < 0.3:
    speed = 0
```

This code uses abs to get the magnitude of the speed only. If the magnitude is below 0.3, it sends a 0 instead. I found 0.3 experimentally, but this may be different on your robot.

Before continuing, you should test this on the robot and check the next section if this behavior did not try to find the distance to the object.

## Troubleshooting

The following are some solutions to try if you aren't getting the follow behavior to work:

- First, check the distance sensor functionality using the *Chapter 8* code. Then, verify the sensor and motor connections.
- If the robot is driving the wrong way, reverse the proportional constant.

There is a problem where this robot may get close to the right place but leave a gap for small distances. As a result, it might not generate a speed large enough to move the motors. In the next section, we'll see why.

## Using the integral to handle small distances

The following graph shows a small error remaining after activity:

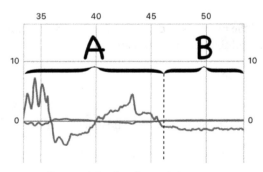

Figure 10.5 – Small remaining error

The preceding figure shows a graph of the error and motor output versus time. The error line varies by up to 10 units. The motor graph is more flat. This line is between 0 and 1 and in the opposite direction, as expected from the proportional experiment. The graph is divided into two portions. Portion A shows the robot moving as we move an object closer and further away, with the motor responding. In portion B, we gradually move the object, creating a small error of -1. This small error results in a speed of 0.1, which is insufficient to move.

What we have is a **steady-state error**. The system has not converged on the set point and will not act further to reduce the error.

This situation is where the integral element is useful. The **integral** is equivalent to an area under a graph. Look at the following graph for an example:

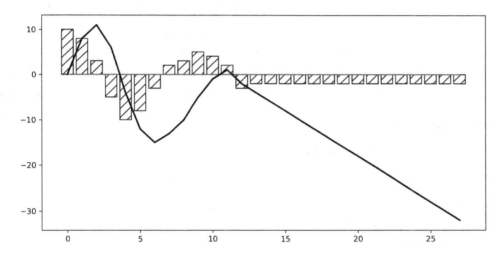

Figure 10.6 – Plotting the integral

This graph has two elements. The bars represent the error value from the previous graph, approximated as discrete time steps. Then, there's the thick line over this, which represents the integral. While the graph varies, the integral varies perhaps a little later. However, when it reaches a steady state, the graph starts to pull downwards continuously.

If we take this and multiply it by another small constant, we can cause the motors to move a little to iron out a steady-state error. We call this the **integral gain**, or ki. The following diagram shows the control system with the integral term:

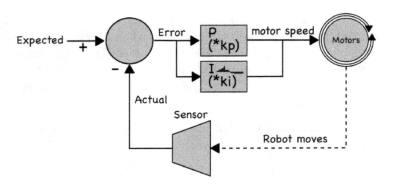

Figure 10.7 – The feedback loop with the integral

The preceding diagram shows the feedback loop control system with the integral added – this box shows $I * ki$, the integral multiplied by an integral gain. It feeds into the motor speed, which is added to the proportional output. The integral is given the same error term as the proportional element.

The integral will store the area of the error graph seen until this point. This would be the area under a graph. How do we calculate the area? We can take each error and multiply it by the time interval during which it was produced. Adding this to a running total represents a good approximation of the area.

The following code extends the previous example.

Let's update the PController code so that it's a PIController class:

```
class PIController:
    def __init__(self, kp, ki):
        self.kp = kp
        self.ki = ki
        self.integral = 0
```

In the preceding code, we added ki as an integral gain to scale our integral. We also store an integral total, starting at 0.

> **Important Note**
>
> Prefixing gain constants like these with $k$ is due to them being treated as *constants* by the PID algorithm, but the outer code can tune these. So, mathematically, this is a constant, but not a constant in the code.

Then, we must change the `calculate` method:

```
def calculate(self, error, dt):
    self.integral += error * dt
    return self.kp * error + self.ki * self.integral
```

`calculate` now adds the error to the integral, which means the integral will continue moving in the error direction. However, we will multiply this by the elapsed time (or delta time, `dt`), so that a longer interval between measurements will result in a larger area. The last line multiplies the current integral by the integral gain.

We'll need to set the integral constant when we create the controller:

```
distance_controller = PIController(-0.19, -0.005)
```

The values of -0.19 and -0.005 work for my robot. We will learn how to tune these later in the chapter. The integral constant should be negative in this case and small. Larger values will cause overshoots, and the system will go back and forth (oscillate).

We can now alter the main loop:

```
prev_time = time.monotonic()
while True:
  if robot.right_distance.data_ready:
    distance = robot.right_distance.distance
    error = distance_set_point - distance
    current_time = time.monotonic()
    speed = distance_controller.calculate(error, current_time - prev_time)
    prev_time = current_time
    if abs(speed) < 0.3:
      speed = 0
    uart.write(f"{error},{speed},"
      f"{distance_controller.integral}\n".encode())
    robot.set_left(speed)
    robot.set_right(speed)
    robot.right_distance.clear_interrupt()
    time.sleep(0.05)
```

The highlighted changes start with the time delta calculations. We keep a previous time (`prev_time`) so that we can subtract this from `current_time` later and feed this time difference into the controller with the error. Next, we store `current_time` in `prev_time` so that we are ready for the next loop. The `time.monotonic` function provides a time in seconds with fractions, guaranteeing that subsequent calls cannot return a lower value.

We also send the current integral value to the UART to plot it. Send this code to the robot:

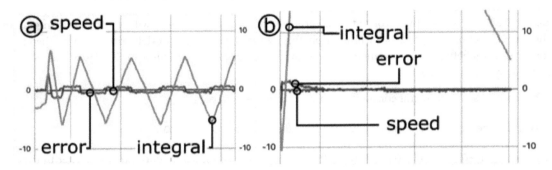

Figure 10.8 – Oscillation versus smaller integral

In *Figure 10.8 (a)* the screenshot shows an error with a large integral gain. This rise makes the motors move and carry on while the system overshoots, until the integral flips the other way - this causes oscillation. In *Figure 10.8 (b)* the integral gain is much lower, so the integral must reach a higher value before it creates a change. The motors respond to the proportional change first, but the integral makes a slight adjustment later. This setting may still oscillate a tiny amount, but the oscillations will be slow and subtle.

Note that when the robot corrects the steady state and the error reaches zero, this doesn't reduce the integral.

> **Important Note**
>
> Stop the code (or robot) if the wheels/motors are not moving the robot forward, and restart the code when the wheels are in contact with the floor or motors are turned on (they are on battery power only). Otherwise, the system can build up a large integral and will ram the next object placed in front of it. This problem is known as **integral wind-up** and can be a big problem if you do not account for it.

Some overshoot here is inevitable, but the system is reacting to steady-state errors. This PI controller will work, but if you move the target object quickly, you may still be able to induce oscillation. How can we dampen that? We'll see how in the next section.

## Dealing with oscillations using the derivative

Oscillations are made due to sudden state changes, creating a large reacting output, which can cause a large change of state in the opposite direction. This may repeat on either side of the set point. The state change is equivalent to a slope on a graph at any point.

The following graph shows a varying value, with the slopes marked at a few points:

Figure 10.9 – Slopes on a graph

The preceding graph shows an error PID response settling. Along the graph are dots, with dashed lines showing the slope. The derivative represents this slope at any point.

The essential factor is that when there is a steep change in the error, there is a steeper slope. This slope value is the **derivative**. It represents the rate of change in our system. Using this value with a gain can reduce overshoot, dampening the system. So, we could add this along with the PI components and dampen the movement a little.

However, before we do that, there is another issue: the output from the distance sensors is a little noisy. The following zoomed-in view shows noise on the sensor when the robot is not moving:

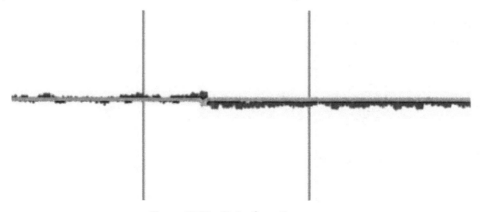

Figure 10.10 – Noise from the sensor

The preceding screenshot is enlarged to show the sensor data noise. One of the lines is frequently going above or below the line. This noise is slight but makes many fast changes in the slope.

Using the derivative directly with this noise may cause our robot to produce large movements. To reduce this, we will use a **low pass filter** in front of the integral. This only allows sustained error movements in, filtering out the constant jiggle of noise.

Let's see this as a feedback control diagram:

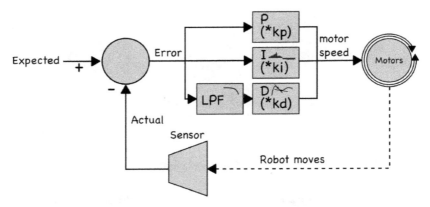

Figure 10.11 – Feedback flow for the PID controller

*Figure 10.11* extends *Figure 10.7* by adding two additional blocks. First, the error splits through a block labeled LPF – this is a low pass filter, with a stylized frequency graph showing how this tails off for higher frequencies. The LPF block feeds into the next block – the derivative multiplied by a kd derivative gain. This block has a stylized graph showing the slope lines. The derivative output is then added to the other PI outputs to make the motor speed signal.

Now that we have seen how this operates, let's modify the code, starting with the PID class:

```
class PIDController:
    def __init__(self, kp, ki, kd, d_filter_gain=0.1):
        self.kp = kp
        self.ki = ki
        self.kd = kd
        self.d_filter_gain = d_filter_gain
        self.integral = 0
        self.error_prev = 0
```

Here, we add two additional construction parameters to the `PIDController` class – a derivative gain, `kd`, and a derivative filter gain. We also store a previous error to calculate the difference from the current error. The `calculate` function also changes:

```
def calculate(self, error, dt):
    self.integral += error * dt
    difference = (error - self.error_prev) * self.d_filter_
gain
    self.error_prev += difference
    self.derivative = difference / dt
    return self.kp * error + self.ki * self.integral +
self.kd * self.derivative
```

We need the difference between the current error and the previous error. We multiply this difference by the filter gain, and add this onto the previous error. This means that the error changes by a smoothed out amount and sets us up for the next calculation.

The next line divides this difference by the change in time, to get the rate of change. We store this in `self.derivative` so that we can graph this term later. Finally, we multiply `self.derivative` by the d gain (`kd`) and add this to the calculations.

The PID controller is now complete. We can make this reusable by moving the `PIDController` class into a `pid_controller.py` file.

The `code.py` file can use `pid_controller` like this:

```
import time
import board
import busio
import robot
from pid_controller import PIDController
uart = busio.UART(board.GP12, board.GP13, baudrate=9600)
robot.right_distance.distance_mode = 1
robot.right_distance.start_ranging()
distance_set_point = 10
distance_controller = PIDController(-0.19, -0.008, -0.2)
prev_time = time.monotonic()
while True:
    if robot.right_distance.data_ready:
        distance = robot.right_distance.distance
        error = distance_set_point - distance
```

```
    current_time = time.monotonic()
    speed = distance_controller.calculate(error, current_time -
prev_time)
    prev_time = current_time
    if abs(speed) < 0.35:
      speed = 0
    uart.write(f"{error},{speed}, {distance_controller.
integral}, {distance_controller.derivative}\n".encode())
    robot.set_left(speed)
    robot.set_right(speed)
    robot.right_distance.clear_interrupt()
    time.sleep(0.05)
```

The highlighted parts of the code show how this changed from the previous code.py example. First, we imported the PIDController class. Then, we swapped our use of PIController for PIDController. PID tuning parameters are tuned as we expand on this code. Adding additional terms, such as the derivative, will make tuning the others necessary.

We send the error and time difference into the controller calculation to get the speed, then send the derivative to the UART to see what our robot is doing and tune the parameters.

With that, we've built a PID controller and used it to keep a certain amount of distance from an object. We'll make this into a more dynamic example in the next section.

## Using PID to follow a wall

Driving along a wall using the PID algorithm requires a little more coordination. Let's visualize the problem with a diagram:

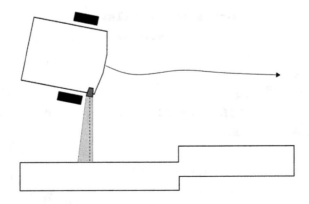

Figure 10.12 – The robot following a wall

*Figure 10.12* shows how our robot will follow a wall. First, the robot drives forward in the direction shown by the solid line with an arrow. We have turned the sensor out so that it can detect the wall in its cone (these distance sensors cover around 20 degrees). Based on the return of a close object (shown as a dashed line), the robot will adjust its heading to try and keep a constant distance. When the robot faces the wall, it will curve outward, and it may overshoot, but also, there is a step change in the wall, so the robot will adjust its path and straighten up.

We have a few issues. First, we have put the motors a little forward, and turning the sensor with the current placement would have the wheel in the path of the sensor, so we'll need to move them back. Then, we'll need to make that 90-degree sensor turn. After, we'll need to use the PID to determine how much to change the robot's heading, a deflection, as the robot drives forward.

Let's start by making some changes to move the sensor.

## Changing the sensor's placement

We will need to make some changes to our CAD sketches, then use a drawing to drill some holes. Note that with the front of the robot facing down in the CAD sketches, the left-hand side of the CAD sketch is for the robot's right-hand side.

The following FreeCAD screenshots show how to make the CAD changes:

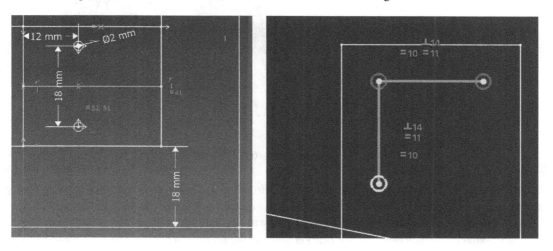

Figure 10.13 – FreeCAD sketches for turning the sensor sideways

The left-hand screenshot of *Figure 10.13* shows how we move the motor back in the `UpperParts` sketch. First, delete the horizontal constraint between the bottom corner of the motor (shown on the left) and the breadboard. Then, add an 18 mm vertical dimension between the same two points. Moving the motor back 18 mm gives ample space for the sensor, but it will also mean we can reuse one of the existing holes since the distance between them is 18 mm. In addition, the two motors have symmetry constraints so that they will move together. Now, you can close this sketch.

The right-hand side of *Figure 10.13* is a screenshot of the sketch for an additional front sensor hole at 90 degrees. Import the existing distance sensor holes as external geometry. Then, use construction lines to constrain the distance between this hole and the top outer hole so that it is equal to the distance between the initial distance sensor holes. Add an equals constraint on the circles and a perpendicular constraint between the lines. You can make a similar circle on the other side, although we will only use one sensor for this demonstration.

The following figure shows the CAD drawing and parts fitted in new positions on the robot:

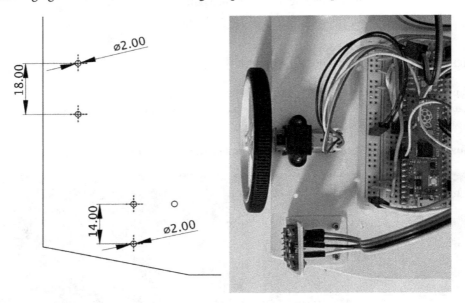

Figure 10.14 – The drawing and result of this robot change

The left-hand side of the preceding figure shows the drawing. I've marked the new holes, using the existing ones as a reference. You will need to detach the motors and sensors to do this – unbolt them but leave the wiring connected and put them carefully to one side while doing this. After removing the motors, attach one side of the motor bracket and use the hole on the other side of the bracket to mark the new hole.

With the distance sensor, remove the inner bolt, then turn it to face out 90 degrees. You can then use the inner bolt hole to mark where to drill.

The right-hand side of the preceding figure shows the motor and sensor moved into position. Ensure that the connections are still correct.

This robot is ready for us to write wall-following code on it. Let's see how.

## Wall-following code

Wall following is an extension of the code we've already been working on, with a few key differences. First, our robot will drive at a constant speed, but depending on the sensor feedback, it will steer closer to/farther from the wall. It can do this by adding a PID output to the speed of one side and subtracting it from another.

Start by taking a copy of the previous code example. We'll make changes to code.py. The imports and setup stay the same:

```
import time
import board
import busio
import robot
from pid_controller import PIDController
uart = busio.UART(board.GP12, board.GP13, baudrate=9600)
robot.right_distance.distance_mode = 1
robot.right_distance.start_ranging()
```

However, along with the other settings, we must add a base speed for the robot to drive. The distance set point should also be further out:

```
speed = 0.7
distance_set_point = 15
distance_controller = PIDController(0.05, 0.0, 0.0)
```

We have set up the PID here so that you can tweak its settings.

The main loop gets the sensor data and error in the same way:

```
prev_time = time.monotonic()
while True:
    if robot.right_distance.data_ready:
        distance = robot.right_distance.distance
        error = distance_set_point - distance
```

When we calculate the PID, we now store it in deflection, describing how fast we will turn:

```
        current_time = time.monotonic()
        deflection = distance_controller.calculate(error, current_
time - prev_time)
        prev_time = current_time
```

We write this to the UART so that we can plot and debug it:

```
uart.write(f"{error},{deflection}\n".encode())
```

The code uses this deflection by adding it to the right motor speed and subtracting it from the left. This deflection should pull us toward the set point, a distance from the wall, while driving forward:

```
robot.set_left(speed - deflection)
robot.set_right(speed + deflection)
```

Finish by resetting the sensor and sleeping a bit before looping:

```
robot.right_distance.clear_interrupt()
time.sleep(0.05)
```

You can send this to the robot and start it near a wall, and it may try to follow it. I've found a few boxes in the middle of the room good as it can drive around them; however, it will likely struggle with anything concave.

The result of this may be unstable, so we'll need to tune this PID. However, even if unstable, it should try to follow the wall, even if it collides.

## Troubleshooting

Try these steps if the robot isn't following or starting:

- First, check the distance sensor wiring; now that you've moved it, the wires may have been dislodged. *Chapter 8, Sensing Distances to Detect Objects with Pico*, contains guides on sensor wiring if you need to check this.

- Ensure you have a fresh set of batteries – this won't work well on low batteries.

- If the robot is spinning, put it closer to the wall, and if it is reacting too hard, bring down the P value. We'll tune this more shortly.

You should now have a robot that is kind of following the wall but may be quite unstable or crashing somewhat. To make this work well, we'll need to tune the PID.

# PID tuning – using graphs to tune the PID

The PID algorithm is great for responding to sensor input, adjusting for constant errors, and dampening out overcompensation with the derivative. The only problem is that getting these three values right is tricky. How you tune a PID depends on the system. In our case, for wall following, much of this will be on how the robot feels in the situation. This method works for small robots, but there are formal methods that require a mathematical model of the system.

Depending on the situation, we may only use one constant, but for this, we'll try to use all three. It's also a good practice to change only one gain constant at a time.

We already have graphing for our system on the phone. While we are starting, we'll only show the error and response; we can add other components as needed. The derivative and integral components can dwarf the proportional component, and the Bluefruit app does not allow plots with different scales on the same graph.

We are also going to add some controls other than graphing. Otherwise, you'll be plugging the robot in to reprogram a lot! So, let's start with some basic controls.

## Controlling motor speed

First, we'll need the ability to start and stop our motors. Currently, our system starts with motors live and running. Turning the motors off or slowing things down will make tuning far less frustrating.

The motor speeds are down to the PWM system. The current PWM settings on our robot have a frequency of 500 Hz. This is a little high for DC motors, which means they can stall (stop) at low speeds. So, we'll reduce the motor PWM frequency. In robot.py, make the highlighted changes on the matching lines:

```
motor_A1 = pwmio.PWMOut(board.GP17, frequency=100)
motor_A2 = pwmio.PWMOut(board.GP16, frequency=100)
motor_B1 = pwmio.PWMOut(board.GP18, frequency=100)
motor_B2 = pwmio.PWMOut(board.GP19, frequency=100)
```

In code.py, we will modify our system to allow motors to be turned off:

```
prev_time = time.monotonic()
motors_active = False
while True:
  if robot.right_distance.data_ready:
    distance = robot.right_distance.distance
    error = distance_set_point - distance
    current_time = time.monotonic()
    deflection = distance_controller.calculate(error, current_
time - prev_time)
    prev_time = current_time
    uart.write(f"{error},{deflection}\n".encode())
    if motors_active:
      robot.set_left(speed - deflection)
      robot.set_right(speed + deflection)
```

```
robot.right_distance.clear_interrupt()
time.sleep(0.05)
```

Then, we'll want to add UART control:

```
if uart.in_waiting:
    command = uart.readline().decode().strip()
```

`if` checks for waiting commands; if we have one, we read in a line of input and decode it. `strip` removes the line end character.

Now, we can start checking for commands:

```
if command.startswith("M"):
    speed = float(command[1:])
```

This code means we can send the robot an instruction such as M0.7 via the Bluefruit UART app, and the straight motor speed will be 0.7.

We also want to be able to activate/deactivate the motors:

```
elif command == "G":
    motors_active = not motors_active
    robot.set_left(0)
    robot.set_right(0)
    distance_controller.integral = 0
```

We can stop or start the motors by sending the G instruction to the robot. This handler toggles the `motors_active` variable. It will always stop the motors. The next loop cycle will turn them on only if it's active, ensuring they stop.

Finally this resets the integral to avoid integral wind up while the motors are not running. Having the integral running while the robot cannot act to counter it is known as **integral wind-up** and can cause big problems.

We can test this with the existing proportional and send it to the robot. You should be able to test the instructions from the UART panel in the Bluefruit app:

- G -> Enable/disable robot motors: think Go/Stop.

- M<speed> -> Set the motor speed: this should be between 0.3 and 1.0. It will likely not move under 0.3. This will behave incorrectly at negative speeds, and the motor code truncates values above 1.0.

Now that you have control of the robot, let's figure out how to tweak the proportional component.

## The proportional component

The proportional gain constant, `pk`, is the starting point and is usually dominant in a system. Multiplying the error by `pk` will create the most immediate reaction to a sensor.

We can start by adding code to change the proportional constant:

```
elif command.startswith("P"):
    distance_controller.kp = float(command[1:])
```

This code checks for the input, `P<proportional gain>`, and uses it in the same way as the motor speed. When you type a value, the proportional constant is updated. So, let's review the state of the robot by adding an instruction to inspect these:

```
elif command.startswith("?"):
    uart.write(f"P{distance_controller.kp:.3f}\n".encode())
    uart.write(f"I{distance_controller.ki:.3f}\n".encode())
    uart.write(f"D{distance_controller.kd:.3f}\n".encode())
    uart.write(f"M{speed:.1f}\n".encode())
    time.sleep(3)
```

This handler for `?` will print the PID constants and speed settings. The `.3f` annotation encodes the value as a number with 3 decimal places. Because the robot usually outputs the graph values via UART, we pause for 3 seconds here. Do not use this command with the motors active.

When we send this to the robot, we have two additional control abilities:

- `P0.045` -> Set the proportional gain value to `0.045`.
- `?` -> Print the current state:

```
P0.045
I0.000
D0.000
M0.7
```

Send this to the robot. Before we engage the motors (`G`), let's consider a good proportional value. For this example, we expect an error between 5 and -5, giving us a range of 10 cm. Corners will go beyond this and set the turning at full (saturation, maximum value). With a speed of 0.7, to turn back, we would need an output of -1.4. We can divide 1.4 by 10, giving us a starting guess of 0.14. Send `P0.14` to set this and `G` to start the robot moving.

This setting is responding but has some oscillation. Use G to stop it. We will use trial and error here by halving the P value and trying again. We can test this and use two rules:

- If the robot is oscillating too much, divide the value by 2.
- If the robot responds too slowly, multiply by 1.5 (the value between here and the previous one is high).

This method lets us home in on a value. When you find values that you like, you can put them back into code.py to keep for later. Use ? to see the last P setting you had.

The following screenshot shows how this looks on the graph output:

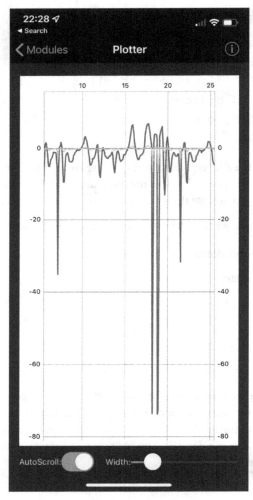

Figure 10.15 – Proportional response

The preceding phone screenshot shows a graph with only a proportional response. The X axis specifies the time in seconds. The Y axis is the distance in cm for the error, but more arbitrary for the output value. The error has a higher range in tens of cms. The robot is driving around some obstacles. The huge swings are when it reaches a corner and drives around it. The blue line shows the PID output and the deflection. The deflection scale is usually below 1, but the large swings will take it above that. Motor speeds are clamped between -1.0 and 1.0, so those large corners will saturate the motors, making the robot turn on the spot.

In summary, a high proportional gain will make a system respond quickly, but it will overshoot and may even start to oscillate. Too low, and it will not respond fast enough.

A proportional-only system can oversteer when there are larger changes. We want to damp this out. We'll adjust the D component to deal with this.

## Adjusting the derivative gain

The derivative component allows the robot to deal with large changes in the error and react by either damping or pulling harder if there's a sudden change, such as oversteering, running out of wall (a corner), or finding a step change in the wall.

Let's modify our code to make it easier to work with the derivative. In code.py, add , {distance_controller.derivative} to the uart.write line after we calculate the deflection:

```
    deflection = distance_controller.calculate(error, current_
time - prev_time)
    prev_time = current_time
    uart.write(f"{error},{deflection},"
        f"{distance_controller.derivative}\n".encode())
```

We also need to control the derivative. Add the following after the P command handler:

```
    elif command.startswith("D"):
        distance_controller.kd = float(command[1:])
```

As you can see, this sets a simple pattern, the same as P.

Now, we can start this with half the P value, D0.035, which results in fewer collisions with occasional bouncing. The following screenshot shows the robot going around boxes:

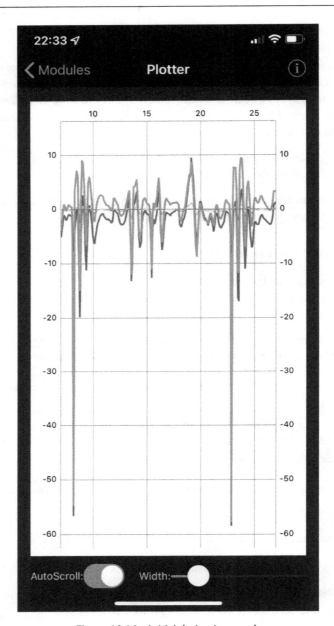

Figure 10.16 – Initial derivative graph

The preceding graph shows the derivative in orange, overlaid on the error in red, and the output in blue. Note that the derivative spikes are very large but can sometimes be the opposite sign of the error allowing it to dampen changes. The largest spikes are still walls. Between 10 and 15 seconds, the derivative dampens the proportional output.

A sudden distance increase, such as the end of a wall, can cause strong oscillation, as shown in the following graph:

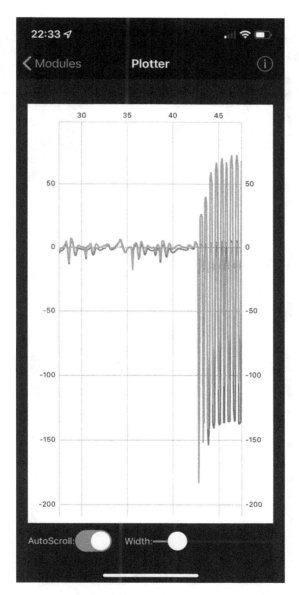

Figure 10.17 – Strong derivative oscillations

In the preceding screenshot, at around 42 seconds, the robot encounters a concave step. That step caused the derivative to overshoot, pushing the robot right out. However, this put the sensor past the obstacle and detected a long-range distance, going rapidly from too close to too far. The derivative swings strongly back the other way – time to stop with G.

We can roughly half the value again to 0.017 for another attempt, but it will still have the same issue. We need to be more aggressive with reducing D than we were with P. We can divide the proportional constant by 10 (0.0035) for a more stable robot. The resulting plot looks as follows:

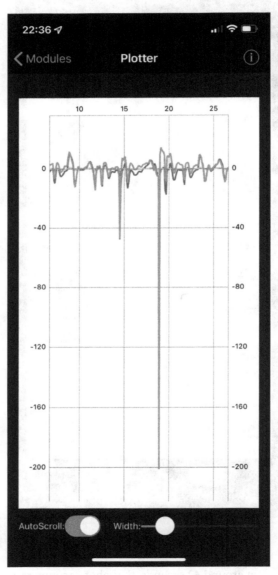

Figure 10.18 – The tuned derivative graph

The preceding screenshot shows a graph with the same properties as *Figure 10.17* but using a 0.0035 setting. While the corners still show up as large spikes, the blue output graph is now calmer, with the derivative dampening the output. The robot will now be driving quite smoothly.

We may need a slightly different scenario to tune the integral.

## Tuning the integral

In an environment with many step changes, such as driving around boxes in a room, the integral will not play as much of a part. Not every problem requires all three parts of the PID, and in this case, the integral may not be particularly suitable. Let's set this nice and slow and find a long straight wall. I had to use a garden path for this (and change some values to suit).

Let's add the integral control and output to the code.py file first. The output is swapping {distance_controller.derivative} for {distance_controller.integral} in the output line.

We can also add this to the control handling:

```
elif command.startswith("I"):
    distance_controller.ki = float(command[1:])
```

This will accept a command such as I0.001 to set the integral. This is also a good starting value. We now have a comprehensive control system.

Send M0.35 to reduce the speed. We can send half the P and D values to match, using ? to see what they were. Send I0.001 and then G to start the robot against a long straight wall.

You should see some large adjustments as the P and D terms settle; then, the I term will settle more slowly. Observe the graph while driving.

For this term, the strategy I use is settling on a small value and slowly incrementing it if the reaction to a constant error is too slow. Like the other terms, making this too high will lead to instability. If the starting point of 0.001 pushes the system into instability, divide it by 10 and slowly increment that.

You now have a strategy to tune PID values on a robot, along with a control system to do so.

## Closing notes on tuning

Tuning a PID will take time. Algorithms such as Ziegler-Nichols can be used to solve this with mathematical modeling and will work in some situations.

Another thing to note is that changes in the system response, such as its turning circle, will alter things. For example, driving a robot on a carpet causes drag on the wheels, making turning slower. If a PID was tuned to work well on carpet, putting the same robot on wooden flooring may cause it to oversteer as its steering effort results in larger changes. Tune a PID for a particular behavior to the environment where you expect it to be operating.

## Summary

The PID controller is a great way to build robot behavior that adjusts to sensor input. In this chapter, you learned what a PID controller and its components are, including where a low-pass filter makes it behave better.

The PID controller allows for dynamic responses but requires a lot of tuning to get it right. You've seen how to add a wireless control method, which is essential for tuning a PID. You've also observed the graphs of PID systems, understanding how they relate to their operations and tuning.

In the next chapter, we will be taking our PID controller and using it with encoders to drive in straight lines.

## Exercises

These exercises will deepen your understanding of the topics discussed in this chapter and make the robot code better:

- Enhance the settings code so that the set point can be adjusted in the same way. `S<set point>` is probably a good idea.
- Modify the command handlers for setting PID values to stop the motors and reset the integral when these values change.
- Try adapting the control code back to the distance control app used earlier in this chapter.
- Try the robot wall-follow when driving on a completely different surface and adjust the PID values to get a smooth wall following.

## Further reading

These study aids will let you read on and dive deeper into the PID algorithm and its quirks:

- Christopher Lam on YouTube has an excellent control theory video course: `https://www.youtube.com/playlist?list=PLxdnSsBqCrrF9KOQRB9ByfB0EUMwnLO9o`. This uses MATLAB and goes into detail about control systems such as PID, tuning them, problems with them, modeling them, and analyzing them. There are at least 30 hours of content that takes a very mathematical approach.
- For a deeper dive into PID control systems, consider *PID-based Practical Digital Control with Raspberry Pi and Arduino Uno* from Elektor Electronics. This book discusses control theory, transform functions, and PID tuning, while providing code and practical examples.
- *PID Control Fundamentals*, by Jens Graf, published via CreateSpace, is a comprehensive look at PID control systems. This provides more detail on each of the components and building PI, PD, and full PID algorithm control systems.

# 11

# Controlling Motion with Encoders on Raspberry Pi Pico

So far in this book, we've added sensors to our robot that can track counts as its wheels turn. We've also looked at a PID algorithm to close the robot control loop. We can combine these concepts to control our motors and wheels more precisely.

This combination will let us maintain a known speed on each motor and correct their relative speeds for a straight line. Encoders with some geometry will let us drive the robot a predetermined distance.

In this chapter, we will cover the following main topics:

- Converting an encoder count into a speed

- Using PID to maintain speed and a straight line

- Driving a known distance

## Technical requirements

For this chapter, you will require the following:

- The robot from *Chapter 10*, *Using the PID Algorithm to Follow Walls*

- The robot, encoder, and PID code from *Chapter 10*

- Around 2 square meters of floor to test the robot on

- Digital calipers

- A PC or laptop with Python 3

- An Android/iOS smartphone with Bluetooth LE

You can find the code for this chapter at `https://github.com/PacktPublishing/Robotics-at-Home-with-Raspberry-Pi-Pico/tree/main/ch-11`.

# Converting an encoder count into a speed

In *Chapter 6, Measuring Movement with Encoders on Raspberry Pi Pico*, we used PIO to retrieve a count from the motor encoding sensors. We ended that chapter by measuring for movement and counting encoder transitions over some time.

In this section, we will relate wheel geometry to the encoder. Then, we will use that to convert encoder counts into a speed or a distance.

## Loose bolts and nuts

Vibration can sometimes cause nuts to drop out – a tiny dab of nail varnish across the nut and thread can reduce this.

## Robot wheel geometry

Calculating the distance traveled by a wheel requires its circumference. Let's start by measuring the diameter of the wheel, as shown:

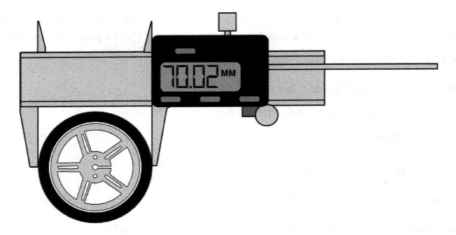

Figure 11.1 – Measuring wheels with calipers

The preceding diagram shows how you can measure wheel diameter with digital calipers. The diameter can be used in our code directly. In `robot.py`, add your measurement rounded to the nearest 0.1 mm:

```
import math
wheel_diameter_mm = 70
wheel_circumference_mm = math.pi * wheel_diameter_mm
```

The preceding code calculates the circumference from the diameter. Each time a wheel makes a complete turn, it will move the wheel circumference in that direction, so we can already convert between wheel revolutions and distance. Next, we need the encoder details.

## Encoder geometry

In *Chapter 6, Measuring Movement with Encoders on Raspberry Pi Pico*, we found the number of poles on the encoder and the number of encoder revolutions per revolution. The N20 built-in magnetic encoders produce 28 edges or state changes for each encoder disk revolution. We then multiply this by the gear ratio 298:1.

We can add these calculations (use your motor gear ratio) to robot.py:

```
gear_ratio = 298
encoder_poles = 28
ticks_per_revolution = encoder_poles * gear_ratio
ticks_to_m = (wheel_circumference_mm / ticks_per_revolution) /
1000
```

We use m and m/s since this puts distances and speeds in the same order as the motor speeds.

Now, we can use these geometry measurements to get a speed.

## Measuring the speed of each wheel

We will calculate the speed of each wheel using the speed triangle from physics:

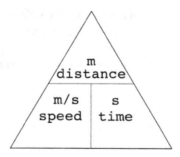

Figure 11.2 – The speed triangle

The triangle in the preceding diagram shows distance over speed and time. We want speed, so we get distance over time by covering speed. We can also see m/s. In our code, we'll need to convert the difference in encoder ticks into a distance in m, then divide that by the time the difference covers:

```
speed = robot.ticks_to_m * (new_position - last_position) /
time
```

We can use this calculation in an app to demonstrate the principle.

### Planning a speed-measurement app

We can build an app to demonstrate this and try different speeds. Using a UART command system will make it interactive.

We'll build the app using `asyncio` – asynchronous input/output. `asyncio` lets us run a few tasks simultaneously on the robot. Most tasks sleep between events, and CircuitPython can run another task during that time. The app must perform the following tasks:

- Measure the encoders, convert them into a speed value, and send this value to the UART.
- Accept the control commands to change settings or drive for a while.
- Stop the robot after a specific time.

The command handlers we'll want for this robot are as follows:

- `M0.7`: Set the motors speed to 0.7
- `T0.3`: Change the measuring time interval to 0.3
- `G3.0`: Go (start moving) for 3 seconds, then stop
- `G`: Stop the robot from moving immediately

With the design created, let's build the app.

### Speed measurement app

We will use the Adafruit **circup** tool to install libraries on Pico. circup can install and update libraries on CircuitPython devices, handling dependencies. See `https://learn.adafruit.com/keep-your-circuitpython-libraries-on-devices-up-to-date-with-circup` for details. Install the `asyncio` library with the following command:

```
circup install asyncio
```

First, we'll move the UART into the `robot.py` file. At the imports, add the following:

```
import busio
uart = busio.UART(board.GP12, board.GP13, baudrate=9600)
```

With that, the UART has been set up for any further examples. We will also add a convenience function at the end of `robot.py`:

```
def send_line(message):
    uart.write(f"{message}\n".encode())
```

This wraps the usual interaction of adding a new line and encoding a message into bytes on the UART.

In a new folder, `measuring_wheel_speeds`, make a new `code.py` file starting with the following imports:

```
import asyncio
import robot
```

These imports are familiar; however, instead of `time`, we are importing `asyncio`.

We can create a `Settings` class to store the current settings, as follows:

```
class Settings:
    speed = 0.7
    time_interval = 0.2
```

This groups the settings; different tasks can access them. `speed` is the motor speed, and `time_interval` is how frequently the code will read the encoders.

Let's learn how to handle the encoders:

```
async def motor_speed_loop():
    left_last, right_last = robot.left_encoder.read(), robot.
right_encoder.read()
```

We use `async def` to turn the function into an async task. We are computing encoder differences, so we keep a last value. We start this with the current encoder reading.

Next, we go into the sensor reading loop, which uses `sleep` to keep that time interval:

```
    while True:
        await asyncio.sleep(Settings.time_interval)
```

This code performs an asynchronous sleep, allowing other tasks to run. We must read both sensors again, getting new values:

```
        left_new, right_new = robot.left_encoder.read(), robot.
right_encoder.read()
        left_speed = robot.ticks_to_m * (left_new - left_last) /
Settings.time_interval
        left_last = left_new
```

We get the speed by subtracting the last value from the new one. Then, we convert that into meters and divide it by the time interval to get a speed in meters per second.

We must also remember to update the last value. We can repeat this for the right sensor:

```
    right_speed = robot.ticks_to_m * (right_new - right_last) /
Settings.time_interval
    right_last = right_new
```

We can finish the motor speed loop by printing the speeds to the UART:

```
    robot.send_line(f"{left_speed:.2f},{right_speed:.2f},0")
```

Notice , 0 at the end of the UART output. We must add this to anchor the graph at 0 so that the plot shows the speed relative to zero.

The next component we'll need is the motor stop task:

```
async def stop_motors_after(seconds):
  await asyncio.sleep(seconds)
  robot.stop()
```

This task will simply wait the given seconds and stop the robot's motors.

We will also need a UART command handler in an async task:

```
async def command_handler():
  while True:
    if robot.uart.in_waiting:
      command = robot.uart.readline().decode().strip()
      if command.startswith("M"):
        Settings.speed = float(command[1:])
      elif command.startswith("T"):
        Settings.time_interval = float(command[1:])
      elif command == "G":
        robot.stop()
      elif command.startswith("G"):
        await asyncio.sleep(5)
        robot.set_left(Settings.speed)
        robot.set_right(Settings.speed)
        asyncio.create_task(
          stop_motors_after(float(command[1:]))
        )
```

Sending G3.0 instructs the robot to wait 5 seconds, drive, and stop after 3 seconds. This 5-second wait allows the user to start the plot tab before the robot starts moving.

The sleep commands now use asyncio.sleep. We also use asyncio.sleep(0) to let other tasks run while waiting for UART input.

Finally, we start the motor speed loop and the command handler, as follows:

```
asyncio.create_task(motor_speed_loop())
asyncio.run(command_handler())
```

> **Is this multithreaded?**
>
> Async code is not multithreaded. Instead, when an asyncio.sleep is used, control is passed to another async block waiting to run. As a result, async code tasks do not access variables simultaneously.

Send this all to the robot. Now, let's see how this works and test it.

### Testing the speed measurement app

I recommend propping the robot on a box for the first test so that its wheels aren't in contact with anything.

Connect to the robot with the Bluefruit LE Connect app and use the UART menu item. You should see zeros. Send G20, which should start the motors moving, and then press the back button and select the plot mode. You will see a graph like the following:

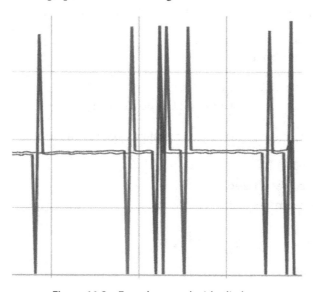

Figure 11.3 – Encoder speed with glitches

The preceding plot shows speed versus time from our robot. The $y$ axis is the speed, while the $x$ axis is the time in seconds. There is a clear 0 line. The graph then shows the two motor speeds. There are glitches – the speed drops to zero and then doubles.

## Fixing the encoder glitches

These glitches are due to an interaction between our read loop and the encoders. Plotting on a UART makes 0.2 s (5 times per second) a good time base. However, our PIO outputs encoder counts as often as they change. The PIO outputs these counts to an RX FIFO queue – see *Chapter 6, Measuring Movement with Encoders on Raspberry Pi Pico*.

The PIO push nowait instruction will write no more data when the FIFO queue is full, but the encoder code continues counting pulses. We can use another asyncio task to read data more frequently from the FIFO queue. In the imports at the top of pio_encoder.py, add the following:

```
import asyncio
```

Add the following method somewhere under QuadratureEncoder:

```
async def poll_loop(self):
    while True:
        await asyncio.sleep(0)
        while self.sm.in_waiting:
            self.sm.readinto(self._buffer)
```

Once started, this will continuously read the data into the buffer as frequently as possible.

Now, we must modify the QuadratureEncoder.__init__ method to create a task for this. Add the highlighted line shown here:

```
self._buffer = array.array("i", [0])
asyncio.create_task(self.poll_loop())
```

The read method can then return the most recent item from the buffer:

```
def read(self):
    if self.reversed:
        return -self._buffer[0]
    else:
        return self._buffer[0]
```

We can now use this encoder code in our async code.

Reupload the `pio_encoder.py` file so that we can try again. Start the motors with G5 and switch to the plot screen; you should see a plot like this:

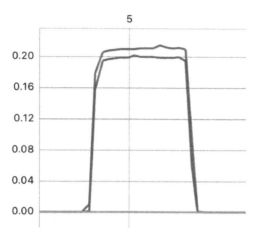

Figure 11.4 – Speed measurement without the glitches

The graph now shows the speed of both motors without the glitches. It is a bit noisy, and one line is slightly higher than the other. One of the motors is also quicker. The robot is moving at around 0.2 m/s. Battery freshness will affect the speed.

If you are not seeing this, please check that the encoders are reporting correctly with the examples provided in *Chapter 6, Measuring Movement with Encoders on Raspberry Pi Pico.*

We can use this measurement to drive a known distance, but it's now clear that the robot won't drive in a straight line like this. So, in the next section, we'll correct the differences between the motors.

# Using PID to maintain speed and a straight line

In this section, we'll learn how to combine the motor distance measurement with a PID controller driving each motor, moving at a particular speed, and keeping the robot straight. Let's start by understanding this system.

## The speed control system

We can set a target speed in meters per second for the robot and compare the converted wheel speeds with it.

The following diagram shows how we'll use this to regulate the robot's driving speed:

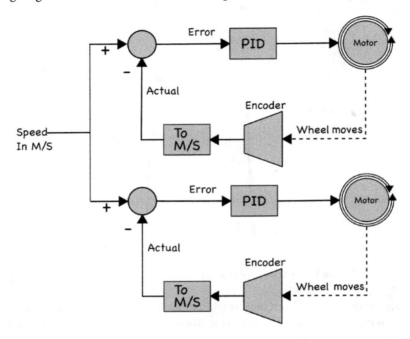

Figure 11.5 – Controlling the speed of two motors

The preceding diagram shows the control system. It starts at the left from a set speed and compares that with the actual speed. The actual speed comes from the encoders, with their ticks converted into m/s. The error is the difference between the speeds.

The error goes into the PID controller, which then produces an acceleration for the motor PWM. The motor power will increase for a positive control signal or decrease for a negative one. This control system repeats for each wheel.

We are building on the motor start/stop with the timer control we used previously. However, with a PID controller, this can cause the integral to wind up and accumulate errors. Let's extend the PIDController class in pid_controller.py so that we can reset this. Make the highlighted change:

```
class PIDController:
    def __init__(self, kp, ki, kd, d_filter_gain=0.1):
        self.kp = kp
        self.ki = ki
        self.kd = kd
        self.d_filter_gain = d_filter_gain
```

```
        self.reset()

    def reset(self):
        self.integral = 0
        self.error_prev = 0
        self.derivative = 0
```

We've moved the initial value settings out to a `reset` method, which we now use in the
`__init__` method.

Now that we understand this concept, we can build the code.

## Speed control code

Let's build our speed controller code. In a new folder, `speed_control`, add a new `code.py` file.
We will start with the regular imports:

```
import asyncio
import time
import robot
import pid_controller
```

We'll add settings that we can adjust to control the system when the program is running:

```
class Settings:
    speed = 0.17
    time_interval = 0.2
    motors_enabled = False
```

We have a `speed` in m/s. This should be close to the speed you measured previously. We also added
a `time_interval` for the loop and specified if the motors should currently be driving.

Next, we must add a `SpeedController` class, which we can use for each wheel system:

```
class SpeedController:
    def __init__(self, encoder, motor_fn):
        self.encoder = encoder
        self.motor_fn = motor_fn
        self.pid = pid_controller.PIDController(3, 0, 1)
        self.reset()
```

We provide each `SpeedController` system with an encoder to read and a motor function (`motor_fn`) to apply control signals. The `SpeedController` creates a `PIDController`. Each wheel will get an independent `PIDController`. This then calls a `reset` function:

```
def reset(self):
    self.last_ticks = self.encoder.read()
    self.pwm = 0
    self.actual_speed = 0
    self.pid.reset()
```

This code puts the first read of the encoder into `last_ticks`, which we'll update when we get a reading. pwm is how much power we will give the motors. We track `actual_speed` so that we can print this value to the UART later. We also reset the PIDs so that any stored integral is gone.

Now, we need a method to update this control system:

```
def update(self, dt):
    current_ticks = self.encoder.read()
    speed_in_ticks = (current_ticks - self.last_ticks) / dt
    self.last_ticks = current_ticks
    self.actual_speed = robot.ticks_to_m * speed_in_ticks
```

This `update` method takes a delta time in seconds. While this might be close to `time_interval`, we need to be accurate when calculating the speed or updating the PID.

The method reads the current encoder value and subtracts the previous encoder reading to get a distance in encoder ticks. To turn this into a speed, we must divide this by time. We must update `self.last_ticks` here for the next cycle. The speed is more useful to us in m/s, so we convert it using `ticks_to_m`.

We can now use this to update the PID and control the motors:

```
    error = Settings.speed - self.actual_speed
    control_signal = self.pid.calculate(error, dt)
    self.pwm += control_signal
    self.motor_fn(self.pwm * Settings.motors_enabled)
```

We multiply the pwm output setting with the `enabled` flag so that the motors will stop if the motors are disabled. We subtract the actual speed from this to get the `error` value.

The code gets `control_signal` from the PID calculation with `error` and delta time. We then use this to accelerate/decelerate pwm, which goes into the motor function.

We use this system to control both motors:

```
left = SpeedController(robot.left_encoder, robot.set_left)
right = SpeedController(robot.right_encoder, robot.set_right)
```

Now, we need an async loop to drive the system:

```
async def motor_speed_loop():
    last_time = time.monotonic()
    while True:
        await asyncio.sleep(Settings.time_interval)
        current_time = time.monotonic()
        dt = current_time - last_time
        last_time = current_time
```

So far, this loop will sleep every interval and update the time, so we have an accurate delta time (dt) value. We can use this to update both sides:

```
        left.update(dt)
        right.update(dt)
        robot.send_line(f" {left.actual_speed:.2f},{Settings.speed
 * Settings.motors_enabled:.2f},0")
```

After updating both sides, we can send the expected speed versus the actual speed to be plotted via UART.

Next, we'll add a modified stop_motors_after async function that updates the motors_enabled flag; it will not call the stop function:

```
async def stop_motors_after(seconds):
    await asyncio.sleep(seconds)
    Settings.motors_enabled = False
```

We want to be able to interact with this. We'll need the command_handler function from the speed measuring app with the highlighted differences:

```
async def command_handler():
    while True:
        if robot.uart.in_waiting:
            command = robot.uart.readline().decode().strip()
            if command.startswith("M"):
                Settings.speed = float(command[1:])
```

```
        elif command.startswith("T"):
            Settings.time_interval = float(command[1:])
        elif command == "G":
            Settings.motors_enabled = False
        elif command.startswith("G"):
            await asyncio.sleep(5)
asyncio.create_task(stop_motors_after(float(command[1:])))
            Settings.motors_enabled = True
            left.reset()
            right.reset()
        elif command.startswith("?"):
            robot.send_line(f"M{Settings.speed:.1f}")
            robot.send_line(f"T{Settings.time_interval:.1f}")
            await asyncio.sleep(3)
        await asyncio.sleep(0)
```

When we send a G<n> command to start the robot moving, we reset left and right, resetting both the previous encoder value and PID integrals. Otherwise, we may have an old encoder setting, and the integral may still hold a value from a previous movement.

All that is left is to start this all up:

```
try:
    motors_task = asyncio.create_task(motor_speed_loop())
    asyncio.run(command_handler())
finally:
    motors_task.cancel()
    robot.stop()
```

This has been wrapped in an additional try/finally block that ensures the movement task is stopped and the robot is stopped if an error occurs.

This code is complete. Send it to the robot along with robot.py, pid_controller.py, and pio_encoder.py.

Ensure the motors are powered on and use the Bluefruit Connect app to send the G10 sequence so that the robot starts moving. I propped the robot up so that its wheels could turn without moving it to initially test this code. This test also lets me keep it connected via USB to see any code errors.

### Speed controller PID tuning

The PID values are likely to need tuning here. The values that worked in my experiments were P: 3, I: 0, and D:1. The P factor will continue accelerating so long as there's a difference, with the D value damping any sudden changes.

I was able to start with a low P value and, using the plot, adjust upward if the overshoot wasn't too great. The following plots show how this system responds:

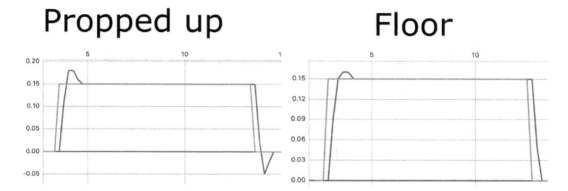

Figure 11.6 – Speed controller response plots

The preceding graphs show the speed controller system. There are two plots – one with the robot propped up so that its wheels have no load and another with the robot on the floor. The orange line shows the set point, which is raised to 0.15 m/s by the code. The blue line is the actual speed measured at one of the motors. The system is better tuned for running on a floor.

Increase the D term to damp the overshoot. Since we are controlling the acceleration of this system, a sustained value is not required for it to keep moving, so the I term can remain at 0.

There may be other troubleshooting issues around motor and encoder connections that you can resolve by going back to the previous examples.

Now, you can control the speeds of two motors simultaneously and get a straight line while practicing PID tuning. We can now build on this to drive in a straight line for a known distance and an expected speed.

## Driving a known distance

We'll need to bring together some of the previous techniques for this. Then, we'll build a variation of the speed controller – a distance controller – so that we can update the distance and let the motor PIDs reach it. This app will use a similar structure, including control and asynchronous tasks, as in the previous example.

## Theory of operation

The following diagram shows the theory of operation for this behavior:

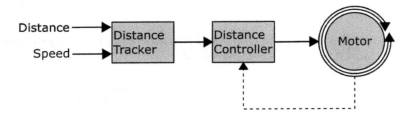

Figure 11.7 – Controlling distance and speed

The preceding diagram shows an overview of this system. The distance tracker tracks distance over time at a given speed, and the distance controller directly controls the motors to try and match a distance. Note the feedback from the motor into the distance controller. We will have one of these per motor.

Let's take a closer look at the distance tracker:

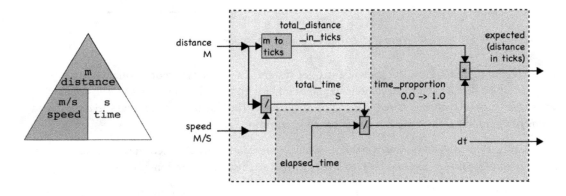

Figure 11.8 – The distance tracker

The left-hand side of the preceding diagram repeats the speed, time, and distance triangle. We have the distance and the speed; to get the time, we need to divide the distance by speed. The right-hand side shows this distance tracker system.

The tracker system must first convert the distance into `total_time` using distance over speed. We must also convert the distance in meters into `total_distance_in_ticks`. These two values will remain constant until we set a new speed or distance. This lighter portion only needs to run when we update the speed or reset the distance.

If we wish to run the system many times, we need to keep track of the current position so that we aren't counting from a position of 0 each time.

When the system runs, the gray portion will run in a loop, and it will be updating the `elapsed_time` since the system was last reset. Dividing `elapsed_time` by `total_time` gives us a proportion of `time_proportion`, which will sweep between 0.0 and 1.0. We multiply this by `total_distance_in_ticks` to get `expected_distance_in_ticks`, tracking the distance the robot should have moved in ticks at any time. Since this component is tracking time, it will also pass along a delta time (`dt`) to the next component.

The next component is the distance controller. Let's take a closer look at this:

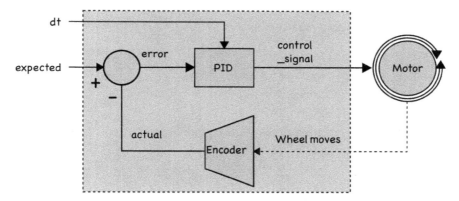

Figure 11.9 – The distance controller

The preceding diagram shows the update method of a `DistanceController` class. `control_signal` directly drives the motor. The `update` method has two inputs –`dt` and the `expected` distance in ticks. We subtract the `actual` distance in encoder ticks from the `expected` distance to get the `error` value, which is used with `dt` to update the `PID`. The output should result in a wheel/motor turning faster if it is behind the expected number of ticks or slower if it's ahead.

We will also use the same graphing and control routines as before, but we will alter the G control handler to specify a distance in meters instead of time in seconds – so, `G0.5` would signal the robot to drive half a meter at the current speed, then stop.

Now, we have enough information to update the code.

## Code to control distance and speed

Let's start with a copy of the previous example code. We will update the specific changed routines.

First, we need to add another number to `robot.py`:

```
m_to_ticks = 1 / ticks_to_m
```

This lets us convert differently so that we can get m from ticks. Let's create the new `code.py` file with the familiar imports:

```
import asyncio
import time
import robot
import pid_controller
```

Now, let's add the `DistanceController` class:

```
class DistanceController:
  def __init__(self, encoder, motor_fn):
    self.encoder = encoder
    self.motor_fn = motor_fn
    self.pid = pid_controller.PIDController(3.25, 0.5, 0.5, d_
filter_gain=1)
    self.start_ticks = self.encoder.read()
    self.pwm = 0
    self.error = 0
```

Here, we initialized the `PID` controller and renamed `last_ticks` to `start_ticks` – the ticks the encoder is at when we start this behavior. We kept `error` so that we can plot it. The code sets `filter_gain` for the derivative to 1 so that the derivative is not too slow in catching up.

Next, we need an `update` method:

```
  def update(self, dt, expected):
    self.actual = self.encoder.read() - self.start_ticks
    self.error = (expected - self.actual) / robot.ticks_per_
revolution
    control_signal = self.pid.calculate(self.error, dt)
    self.motor_fn(control_signal)
```

First, we have an additional `expected` parameter (in ticks). We get an `actual` (in moved ticks) by subtracting `start_ticks` from the current encoder reading. We store it as `self.actual` so that we can graph this.

error is far simpler; subtracting expected from actual gives us the number of encoder ticks we are short or ahead of. However, to scale it down, we must divide by ticks per revolution. This goes into the PID calculation, along with dt.

We use the output of the PID to control the motor.

We've completed the DistanceController code. Next, we need to create the DistanceTracker class. We start by storing the controller's settings:

```
class DistanceTracker:
  def __init__(self):
    self.speed = 0.17
    self.time_interval = 0.2
    self.start_time = time.monotonic()
    self.current_position = 0
    self.total_distance_in_ticks = 0
    self.total_time = 0.1
```

After setting the speed and time_interval fields, we store a start_time to count the time elapsed. We also set up initial values for current_position, total_distance_in_ticks, and total_time. Note that total_time must not be zero as we use it in the division.

We will need to set these values when we update the distance:

```
  def set_distance(self, new_distance):
    self.current_position += self.total_distance_in_ticks
    self.total_distance_in_ticks = robot.m_to_ticks * new_
distance
    self.total_time = max(0.1, abs(new_distance / self.speed))
    self.start_time = time.monotonic()
```

The first thing we must do is add any previous movement to current_position. This means we keep track of the expected position so that the system doesn't accumulate too many errors.

Then, we must calculate a total distance, converting from meters into ticks into total_distance_in_ticks. The code calculates the total time by dividing new_distance by speed. However, since going backward would be a negative speed, we use the abs function to get only a positive time. Also, to avoid that division by zero, we clamp this value to be above 0.1 seconds.

Finally, this resets to a new start_time, from which total_time will be relative.

Now, we can build the `DistanceTracker` loop:

```
async def loop(self):
    left = DistanceController(robot.left_encoder, robot.set_
left)
    right = DistanceController(robot.right_encoder, robot.set_
right)
    last_time = time.monotonic()
```

This code creates two `DistanceController` instances and stores a `last_time` value for `dt` calculations. The next part of the code is all about time:

```
while True:
    await asyncio.sleep(self.time_interval)
    current_time = time.monotonic()
    dt = current_time - last_time
    last_time = current_time
    elapsed_time = current_time - self.start_time
    time_proportion = min(1, elapsed_time / self.total_time)
```

First, we sleep for `time_interval`, then get `current_time`. From this, we can calculate `dt` and `elapsed_time`. We calculate a `time_proportion` between 0 and 1 by dividing `current_time` by `total_time`. This `time_proportion` lets us track where we are in the current motion. Note that we clamp this to a limit of 1 so that the time ratio doesn't multiply to a value larger than the intended distance.

Then, we can multiply `time_proportion` by `total_distance_in_ticks` to get the relative tick position for the robot. As this is a relatively expected position, we add `current_position` again. The `expected` value is an absolute position from when we start the code:

```
    expected = time_proportion * self.total_distance_in_ticks
+ self.current_position
    left.update(dt, expected)
    right.update(dt, expected)
    robot.send_line(f"{expected:.2f},{left.actual:.2f},0")
```

Now, we must update the two `DistanceController` instances and write data to the UART to be plotted.

We can start this part of the system by creating an instance of `DistanceTracker`:

```
distance_tracker = DistanceTracker()
```

We can complete this behavior by creating a UART command handler:

```
async def command_handler():
  while True:
    if robot.uart.in_waiting:
      command = robot.uart.readline().decode().strip()
      if command.startswith("M"):
        distance_tracker.speed = float(command[1:])
      elif command.startswith("T"):
        distance_tracker.time_interval = float(command[1:])
      elif command == "G":
        distance_tracker.set_distance(0)
      elif command.startswith("G"):
        await asyncio.sleep(5)
        distance_tracker.set_distance(float(command[1:]))
    await asyncio.sleep(0)
```

The G<number> command now updates a distance instead of time; stopping the robot sets a new distance of zero.

All that remains is to start the async tasks and handle errors:

```
try:
  motors_task = asyncio.create_task(distance_tracker.loop())
  asyncio.run(command_handler())
finally:
  motors_task.cancel()
  robot.stop()
```

We now have code that tracks a distance at a given speed; however, you will likely need to tune the PID. DistanceController needs to provide enough control_signal to keep up with a changing expected position. There will be a non-zero error value in the update method if it's not completed the motion. This PID system will be dominantly proportional so that the motors keep up with the expected position. The tips at https://pidexplained.com/how-to-tune-a-pid-controller/ help with this tuning.

Note that if you change the motors, the floor type, or the time base, you may need to tune this system again.

You should now have a tuned system to drive a specific distance at a specific speed.

# Summary

In this chapter, you learned how to use the encoder more usefully. You looked at how to use wheel geometry to convert encoder pulses into metric measurements and then used these measurements to measure speed.

Once we can measure speed, we can use a PID to control the speed of each wheel and see significantly less veering.

We could then take this to the next level and drive a specific distance at a specific speed, providing fully controlled motion.

In the next chapter, we will connect an IMU to our robot so that we can measure a compass heading and control the robot's direction.

# Exercises

These exercises will deepen your understanding of the topics covered in this chapter and make the robot's code better:

- All the preceding examples could benefit from the PID modification and printing menu in the UART command handler – consider adding it to them.
- In the distance control, we set the derivative filter gain to 1, disabling it. How does this system behave with other filter gain values?
- Instead of starting motions with the phone app, could you chain some of these movements together? Or even alter the phone G instruction to make a few motions with a single command sequence?

# Further reading

These further study aids will help you learn more and dive deeper into using encoders to control robot motion:

- In *Learn Robotics Programming Second Edition*, *Chapter 11*, *Programming Encoders with Python*, I used simpler encoders but dove into the calculations needed to make specific turns with encoders that could be adapted to the Pico.
- The Arduino-based tutorial at `https://circuitdigest.com/microcontroller-projects/arduino-based-encoder-motor-using-pid-controller` shows how to use a PID controller.
- This Python file at `https://github.com/pimoroni/pimoroni-pico/blob/main/micropython/examples/inventor2040w/motors/position_control.py` from Pimoroni shows a similar Python approach in MicroPython for the Pico.

# 12

# Detecting Orientation with an IMU on Raspberry Pi Pico

Our robot can track how far it's moved, but what about tracking which direction the robot is facing? Or how far it has turned? In this chapter, we will learn about the **Inertial Measurement Unit** (**IMU**), a device that can track the motion of the robot measured against gravity and the Earth's magnetic field.

We'll look at how to select one of these devices, get it connected and then write code for it on our robot using the PID controller to steer the robot based on the IMU data.

In this chapter, we will cover the following main topics:

- What is an IMU and how to choose one
- Connecting the IMU to the robot
- Calibrating and getting readings
- Always face North behavior
- Making a known turn behavior

## Technical requirements

For this chapter, you will require the following:

- The robot from *Chapter 11, Controlling Motion with Encoders on Raspberry Pi Pico*
- The robot, encoder, and PID code from *Chapter 11, Controlling Motion with Encoders on Raspberry Pi Pico*
- A screwdriver, bolts, and stand-offs
- Dupont jumper cables
- A space where strong magnets can be avoided

- A PC or laptop
- An Android/iOS smartphone with Bluetooth LE and the Bluefruit LE Connect app

You can find the code for this chapter at `https://github.com/PacktPublishing/Robotics-at-Home-with-Raspberry-Pi-Pico/tree/main/ch-12`.

# What is an IMU and how to choose one

In this section, we'll look at the components of an IMU and what criteria we used to choose the one used in this robot.

## Components of an IMU

An IMU is a module that can measure movement. It uses multiple sensors to achieve this. In this section, we'll briefly look at each sensor and how they contribute to the whole measurement.

These sensors are made using the **Micro-Electro-Mechanical-Systems (MEMS)** process. They have tiny moving parts embedded into the chips. We can model them mechanically to understand them. These parts sense the movement of parts through their magnetic fields and amplify tiny signals. Let's look at the components.

### *The thermometer*

The mechanical components of an IMU will change size, depending on their temperature. These tiny changes may be enough to change the signals so that the IMU controller can use a temperature measurement to compensate for this.

### *The accelerometer*

An **accelerometer** measures acceleration forces. It measures this as a vector – a direction and a size. The way this is measured is somewhat like a box with a suspended mass, as shown in the following figure:

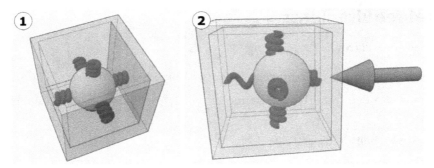

Figure 12.1 – Accelerometer modeled as a mass with springs

The preceding figure shows a mass suspended in a box by springs. When a force acts on the box, the mass retains its inertia and compresses the spring opposing the force's direction. A MEMS accelerometer uses tiny silicon springs and masses and measures the electrical field changes when the mass moves.

While on Earth, gravity pulls the mass down. This system behaves like a force holding the box up, so an accelerometer registers an upward force. We can use this measurement to determine what the downward direction is and sense the tilt of a robot.

The accelerometer vector is an absolute orientation (see the *Absolute and relative sensing* section in *Chapter 6, Measuring Movement With Encoders on Raspberry Pi Pico*) for up. Still, other movements cause noise, so it is usually put through a low pass filter, only changing a setting by a percentage of the actual variation. This filter makes the output slow but stable.

A controller can combine this data with other sensors for a faster and more stable measurement, such as a gyroscope.

### The gyroscope

A **gyroscope** measures the rotation speed of a system, typically in degrees or radians per second in an angle around each axis. A physical gyroscope model, shown as follows, can be used to help illustrate what it does:

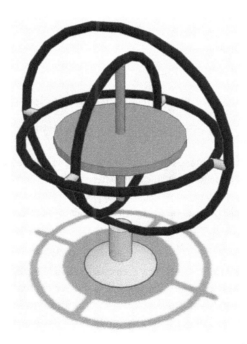

Figure 12.2 – A model of a gyroscope

The preceding figure shows a traditional gyroscope. This gyroscope has a spinning mass wheel in the middle, linked to concentric rings – each ring pivots in one direction – x, y, or z. The net effect is that when you move the handle, the spinning disk preserves its orientation. Sensors placed at the pivots would detect how much the system has rotated in each direction.

The MEMS version uses a tiny mass that's moved back and forth in one direction. If the orientation is changed, the mass will continue vibrating in the original direction, which will change the electrical fields detected by the sensor. This movement in the original orientation appears to be a force known as the **Coriolis force**. The gyroscope can measure the magnitude of this force.

It's essential to understand the directions of the gyroscope and how the measurements relate to time. See the following diagram:

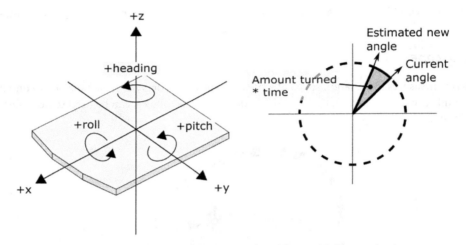

Figure 12.3 – Gyroscope directions and problems with integrating it

The left-hand side of the preceding diagram shows the three gyroscope rotations and the coordinate systems that the IMU uses. This coordinate system takes the robot into account. Traditionally, the front of the robot is in the positive X direction. Straight up is positive Z, and to the robot's left is positive Y. Rotation around the z axis is known as **heading** (also known as yaw), rotation around the y axis is pitch, and rotation around the x axis is roll. This combination of three angles to determine orientation is known as **Euler** (pronounced *oil-er* angles).

The right-hand side of the diagram shows how a controller can use gyroscope data – it represents a relative change in angle or a speed in angle change over time. We can convert this into a new angle, but that is estimated. Multiplying this by time and adding them can give us a whole rotation relative to the starting position, but this will magnify any estimation or reading errors.

A controller should combine this with other sensors, where the gyroscope can provide a fast relative measurement, and the other sensor can provide a slower absolute measurement. However, the accelerometer cannot measure the heading. For that, we need a magnetometer.

## The magnetometer

A **magnetometer** is sensitive to magnetic fields. It passes electricity through a material that creates current when exposed to a magnetic field, as shown here:

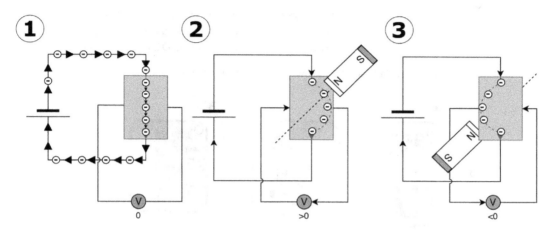

Figure 12.4 – Diagram of a hall-effect sensor

The preceding figure shows an example of detecting magnetic fields in action:

1.  The circuit passes an electric current from a source through a conducting plate (the gray rectangle). The arrows show the flow of electrons (negative charge carriers) moving around the circuit from the top of the plate downwards. The small circle with a V measures electrical flow across the sides of the plate. Currently, this reads 0 as the flow is straight down.

2.  When we place a magnet above the plate, it deflects electrons to one side. This deflection creates a small electric flow through the measuring circuit and will show a reading above 0 on the sensor.

3.  When we place a magnet below the plate, this deflects the electrons to the other side, creating a small sensor flow in the other direction, and show a reading below 0.

This sensor method is known as the **Hall effect**. By measuring three plates, you can sense magnetic fields in three dimensions.

The great thing is that we can use this to sense the Earth's magnetic field and magnetic North, although it can be deflected by magnets in objects around the magnetometer. Being able to sense magnetic North makes it a great way to sense heading.

Because it is subject to other magnets, it can be noisy and may need a low pass filter, but you can pair it with the gyroscope the same way the accelerometer is paired.

Now that you've seen the four sensor types that go into an IMU, we can look at how we choose one.

## Choosing an IMU module

There are several IMU devices on the market. The first thing to note is that you want a module or breakout, not a bare chip (at least not yet).

The number of directions/movements an IMU system can sense is known as **degrees of freedom**, or **DOF**. A system with all three types of sensors is known as a 9-DOF because each can produce three axes worth of information. The temperature sensor isn't counted in this DOF count usually.

These modules come in a few flavors. The following diagram illustrates these flavors:

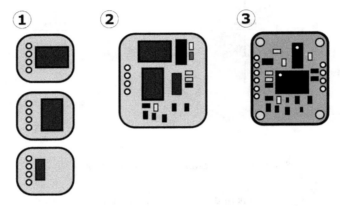

Figure 12.5 – IMU module integration levels

The preceding diagram shows three different IMU module integration levels. They are as follows:

1.  The simplest has only one of the sensors, and you need to buy three to get the complete orientation. In addition, they do not have much helper circuitry.

2.  Some modules integrate three or four separate sensor chips to give the full 9-DOF and could be suitable. These have some support circuitry but might not have a controller.

3.  The devices we will focus on are those based around a single chip that combines the sensors. These come as modules that integrate power and any additional required components (such as small resistors and capacitors). These have onboard controllers.

They can also use a few different data buses to communicate. UART and SPI tie up a whole set of pins; however, I2C allows the device to share a bus with other sensors, so we'll favor I2C devices.

The next factor in integration is how much calculation the device's controller can do (if any). The algorithms to combine all three sensors and account for calibration, along with temperature, are complicated. Some devices can perform this on board, and some require it on another controller or require specialist code to be uploaded to activate the calculation (such as MPU and ICM TDK series). We will also favor modules that can do the calculation on board.

The BNO055 module is a good fit for these requirements. The BNO055 combines all 9-DOF, a temperature sensor, works over I2C, and has calculations and calibration on board, saving us writing that code on our Raspberry Pi Pico. They are also widely available, with Adafruit selling them in two variations, and there's direct CircuitPython support for them.

Let's look at how we can use a BNO055 module with our robot.

# Connecting the IMU to the robot

Installing the BNO055 requires performing a few steps. In this section, we'll prepare the module, attach it to the robot rigidly, wire the part into the circuit, and then use some simple code to test that it is responding.

## Preparing the BNO055

The BNO055 from Adafruit comes without the headers attached. You'll need to solder the headers in, as we have done previously. Adafruit has a guide for this at `https://learn.adafruit.com/adafruit-bno055-absolute-orientation-sensor/assembly`.

For this robot, you should solder this part with the headers facing up from the component side.

## Attaching the BNO055

To attach the part to the robot, see the following diagram:

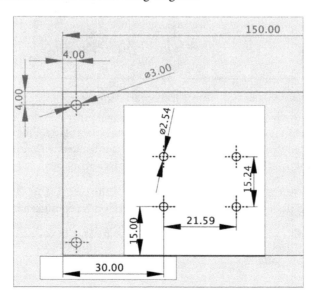

Figure 12.6 – Drawing of the shelf with additional holes for the BNO055 module

You will need to attach the IMU rigidly to the robot, so the velcro pad will not be sufficient. Stand-offs will make a suitable attachment here. The preceding figure shows where to make some 2.5 mm holes in the shelf in the highlighted area. You can insert M2 (or M2.5 if you have them) stand-offs to attach it. You can use stand-offs to gain some separation between the IMU and the metal or magnetic parts of the robot.

You may need to adapt this to the BNO breakout you have. The following figure shows the part I am using and the orientation it should be in:

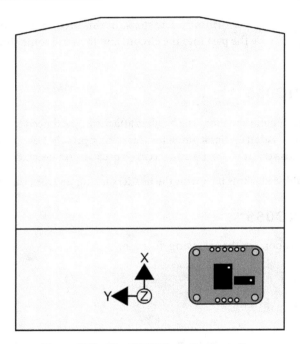

Figure 12.7 – The BNO055 part's orientation

I've made the holes so that they suit the part shown in the preceding figure. This figure shows the part with the robot chassis as a reference, with the x, y, and z axes indicated with arrows. The z in a circle means it runs through this diagram, with the upward direction being positive.

We mount the BNO055 so that its x axis faces the front of the robot. This is due to the BNO055 defaulting to the Android phone orientation – a minor quirk of the configuration of this module.

While the orientation of this part matters, it can be compensated for in code.

Let's see how to wire in this part.

## Wiring the BNO055 to Raspberry Pi Pico

We will wire the BNO055 using I2C. See the following circuit diagram for details:

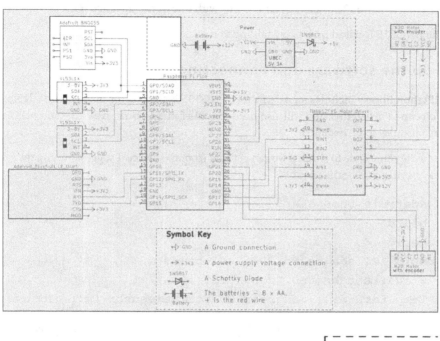

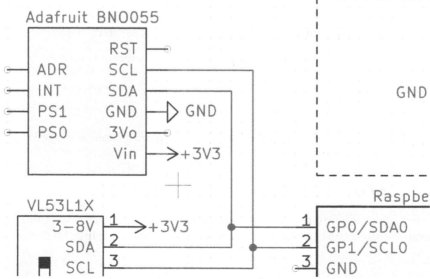

Figure 12.8 – BNO055 I2C wiring

The top part of the preceding diagram shows the whole circuit diagram. Since a lot is happening here, the highlighted region has been zoomed in below, showing the specific wiring. The BNO055 only needs four pins connected. It has power connections for GND and Vin from the 3V3 lines. The SCL and SDA are connected to I2C0, sharing an I2C bus with one of the distance sensors.

Now that the BNO055 is wired in, let's try talking to it.

## Setting up the software and connecting

The CircuitPython library includes an `adafruit_BNO055` module for use with this device. Copy over or use **circup** to install this. You will also need the `adafruit_bus_device` and `adafruit_register` modules to be installed.

You can now write some code to check if we can get data from the device. I suggest putting the following code in `bno_connect/code.py`:

```python
import adafruit_bno055
import board
import busio
i2c = busio.I2C(sda=board.GP0, scl=board.GP1)
sensor = adafruit_bno055.BNO055_I2C(i2c)
print("Temperature: {} degrees C".format(sensor.temperature))
```

The code starts with imports. It then creates an I2C bus with the correct pins and constructs the sensor control object on this bus.

The last line fetches the temperature from the sensor and prints it to serial.

Send this to Raspberry Pi Pico as `code.py`; it should show the temperature on the serial, as follows:

```
code.py output:
Temperature: 21 degrees C
```

Great! You have now obtained data from the IMU. Before exploring more of the available features, let's troubleshoot any problems.

## Troubleshooting

If you don't see the temperature output (or see errors instead), verify the connections carefully. Start by powering down the robot, and carefully check the power goes to 3V3, then that GND goes to ground.

If these look correct, verify the SCL and SDA lines – swapping these lines is a common issue. You can then power the robot again, and you should be able to read the temperature.

With that, you have connected to the IMU. Now, let's learn how to get robot orientation information from this sensor.

# Calibrating and getting readings

When you start up code using an IMU module with a controller, the sensors will not get correct readings. So, the IMU module will need to determine the sensitivity and correct states of the sensor, a process known as **calibration**. First, we need some code; then, we'll need to take the robot through some motions to perform this.

## Calibration code

Let's start with the code. In a file called `imu_calibration/code.py`, add the following:

```
import adafruit_bno055
import board
import busio
import time
i2c = busio.I2C(sda=board.GP0, scl=board.GP1)
sensor = adafruit_bno055.BNO055_I2C(i2c)
```

This code handles importing the module and setting it up. We also import `time` so that we can use it in loops later.

Next, we must check the calibration state of the module:

```
def check_status():
    sys_status, gyro, accel, mag = imu.calibration_status
    print(f"Sys: {sys_status}, Gyro: {gyro}, Accel: {accel}, Mag: {mag}")
    return sys_status == 3
```

This code will print the calibration status for each part of the BNO055. The BNO055 can self-calibrate when turned on; however, the user needs to make motions with it to help. The `calibration_status` register holds what parts of the system you have calibrated. The system status is important for our purposes, but each device has its own status. Each can go from state 0 (uncalibrated) to state 3 (fully calibrated). When you have calibrated them, the system is ready to use. This code will print them out. We'll use this to perform the calibration process motions.

We can check this in a loop:

```
while not check_status():
    time.sleep(0.1)
```

Once we've calibrated it, we can check the data from all the sensors and the controller itself:

```
while True:
    data = {"temperature": sensor.temperature,
            "acceleration": sensor.acceleration,
            "magnetic": sensor.magnetic,
            "gyro": sensor.gyro,
            "euler": sensor.euler}
    print(data)
    time.sleep(0.1)
```

This code will pull all the sensor data and ask the controller to convert the accelerometer, magnetometer, and gyroscope data into **Euler** absolute orientation. The code puts this into a dictionary so that when we print it, it will appear labeled.

We can upload this code and start the calibration process, watching the numbers in the calibration status. Let's use it to calibrate.

## The calibration process

The following movements might look bizarre, but the IMU module is trying to determine the relative motions for each sensor. Then, with the hold postures, it is looking at absolute states. With the magnetometer, there will be offsets and distortions due to the metal on board the robot. The sensor looks for magnetic field changes and the extent in each direction, which it can use to account for the distortions.

Beware of calibrating near strong magnetic fields such as a laptop – they can make the magnetometer calibration incorrect.

Use the following figure to help the IMU complete its calibration:

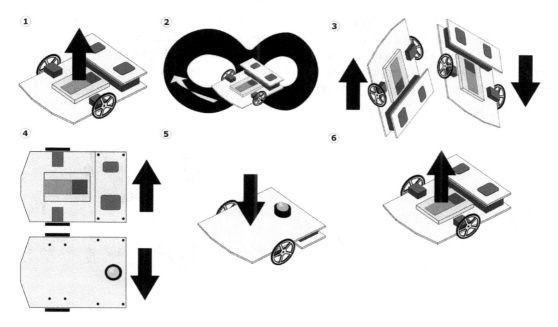

Figure 12.9 – IMU calibration steps

The preceding figure shows the calibration steps. Use a hold time of at least 2 seconds and slow motions for the following steps:

1. Start with the robot on a flat surface and hold. This position will set up the gyroscope.

2. Make a small, slow figure of 8 motion a few times to calibrate the magnetometer.

3. Hold the robot on its back, then on its front.

4. Then, hold the robot on its left, then on its right.

5. Hold the robot upside down; you should see the accelerometer status reach 3.

6. Now, rest it back the right way up. You should be able to see the system status reach 3.

This calibration may take a few attempts and can get stuck on the accelerometer sometimes; however, the experiments can continue if the system status reaches 3 without the accelerometer.

When you have the system status at 3, the demonstration will start printing data from all four sensors and combine sensor data into Euler angles. With the robot standing, roll and pitch should be 0. Turning the robot to face North should set the heading to 0 too.

We can now use this with a PID controller to make the robot always face North.

## Always face North behavior

We'll build a behavior with a heading as a set point for a PID and the IMU Euler heading as feedback. The error value between these will be how far, in degrees, the robot is facing away from the North heading. For example, a heading of 0 should be North – note that you could pick another heading as needed. We will use the PID output to control the motor movements, with the output adding to the speed of one motor and subtracting from the other, producing a turn.

Let's see how this looks as a block diagram:

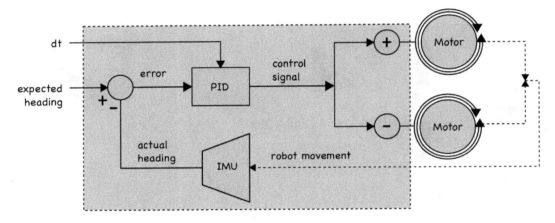

Figure 12.10 – Face North behavior block diagram

The preceding diagram shows the flow of data. The **expected heading** (or target) with the **actual heading** from the IMU are used to calculate the **error**. This error and dt (delta time) are the inputs to the PID. The output from the PID, the **control signal**, is added for one motor and subtracted for the other. The motors then result in robot movement, which causes the IMU heading to change, feeding back into the error value.

We can now use this block diagram to build the code for this behavior.

## CircuitPython code for the face North behavior

We can now build the code for this behavior. We'll start by putting the IMU initialization in robot . py. Add the following to the imports at the top of robot . py:

```
import adafruit_bno055
```

Since the distance sensors already use the I2C bus, we can use the same initialized I2C bus for the BNO055:

```
left_distance = adafruit_vl53l1x.VL53L1X(i2c0)
right_distance = adafruit_vl53l1x.VL53L1X(i2c1)
imu = adafruit_bno055.BNO055_I2C(i2c0)
```

The IMU will be available from `robot.py` once this has loaded. We can also add the `check_status` calibration function to `robot.py` so that we can use that in other behaviors:

```
def check_imu_status():
    sys_status, gyro, accel, mag = imu.calibration_status
    uart.write(f"Sys: {sys_status}, Gyro: {gyro}, Accel: {accel},
Mag: {mag}\n".encode())
    return sys_status == 3
```

The changes to `robot.py` for this section are complete.

We will need a new file for the behavior, which I suggest placing in `face_north/code.py`. We'll start with the imports:

```
import robot
import pid_controller
import asyncio
import time
```

We will then make a controller class for always facing North. It starts by defining the target as 0 for North and setting up a PID controller:

```
class FaceNorthController:
    def __init__(self):
        self.pid = pid_controller.PIDController(0.01, 0.010, 0)
        self.target = 0

    def update(self, dt, angle):
        error = self.target - angle
        if error > 180:
            error -= 360
        elif error < -180:
            error += 360
```

```
        control_signal = self.pid.calculate(error, dt)
        robot.set_left(control_signal)
        robot.set_right(-control_signal)
```

The code here calls the update method with an angle and a delta time (dt). First, it will calculate the error in degrees. The method then checks the error in the range of -180 to 180 degrees. Otherwise, a robot at 359 degrees (-1 degrees from North) will turn a full circle to adjust itself, and any overshoot would result in unusable behavior.

Then, we pass error and dt into the PID calculate method and send the resulting control signals to the motors.

We can now have an async task to manage this controller and read the sensor data in a loop:

```
async def control_loop():
    controller = FaceNorthController()
    last_time = time.monotonic()
    while True:
        await asyncio.sleep(0.1)
        next_time = time.monotonic()
        dt = next_time - last_time
        last_time = next_time
        angle = robot.imu.euler[0]

        controller.update(dt, angle)
        robot.uart.write(f"{angle}, 0\n".encode())
```

control_loop creates an instance of our FaceNorthController shown previously. It sleeps and manages the delta time, dt. Next, it reads the angle from the sensor's euler data and passes this to the update method. Finally, this method logs the angle through Bluetooth so that we can plot it.

Now, we can make our main async function:

```
async def main():
    while not robot.check_imu_status():
        await asyncio.sleep(0.1)
    robot.uart.write("Ready to go!\n".encode())
```

This part will start a calibration/status loop and print it via Bluetooth when the robot is ready. Because we don't want the robot to start trying to drive out of your hands, we will make it wait for a start signal from Bluetooth:

```
while True:
    if robot.uart.in_waiting:
        command = robot.uart.readline().decode().strip()
        if command == "start":
            break
    await asyncio.sleep(0.1)
await control_loop()
```

The user will see **Ready to go** and then need to send `start` to make the robot move. The code then starts the `control_loop` part.

Finally, we can start everything up by starting the `main` task:

```
asyncio.run(main())
```

You should be able to send this to the robot and calibrate it. Then, when you instruct it to start, the robot will turn to face North.

### Troubleshooting

The robot may be turning to an angle that is not North. The common reason for this is that there is a strong magnetic field where you are testing or calibrating the robot. In some situations, I have found that I've had to turn a sensor 90 degrees for it to work.

If the robot is overshooting, try reducing the P value. If it is taking a while to hunt out the actual value, increase the I value a little. I have found that the D value doesn't help in this situation.

Now that we know how to face one way, can we use this to make a fixed turn in any direction? Let's see.

## Making a known turn behavior

The known turn behavior is a variation of the always face North behavior. The idea is to measure the angle at the start of the turn and then make the set point the new intended angle.

We'll make it so that the whole app will accept a difference in the intended angle, offsetting the last intended angle, with the whole app starting based on the robot's current heading. The user can send +30 to turn 30 degrees and -90 to rotate 90 degrees back.

The block diagram is exactly as before, as we only need to manipulate the expected heading. Make a copy of face_north in a folder called known_turn. Let's rename the controller IMUTurnController:

```python
class IMUTurnController:
    def __init__(self):
        self.pid = pid_controller.PIDController(0.01, 0.008, 0)
        self.target = 0
```

The update method doesn't change, as shown here:

```python
def update(self, dt, angle):
    error = self.target - angle
    if error > 180:
        error -= 360
    elif error < -180:
        error += 360
    control_signal = self.pid.calculate(error, dt)
    robot.set_left(control_signal)
    robot.set_right(-control_signal)
```

We will need an additional Bluetooth command_handler for accepting user input for the intended angle. Add the following code:

```python
async def command_handler(turn_controller):
    while True:
        if robot.uart.in_waiting:
            command = robot.uart.readline().decode().strip()
            if command.startswith("-"):
                turn_controller.target -= int(command.lstrip('-'))
            elif command.startswith("+"):
                turn_controller.target += int(command.lstrip('+'))
        await asyncio.sleep(0)
```

This handler sets the target (set point) of turn_controller for dealing with positive and negative number settings.

We can now integrate these into a modified control_loop:

```python
async def control_loop():
    controller = IMUTurnController()
```

```
controller.target = robot.imu.euler[0]
asyncio.create_task(command_handler(controller))
last_time = time.monotonic()
while True:
  await asyncio.sleep(0.1)
  next_time = time.monotonic()
  dt = next_time - last_time
  last_time = next_time
  angle = robot.imu.euler[0]

  controller.update(dt, angle)
  robot.uart.write(f"{angle}, 0\n".encode())
```

This control loop sets the controller target as the current robot's heading instead of 0. It will also create the command handler async task with the controller as a parameter.

The loop is the same as what we saw previously.

The `main` method for this gets to be much simpler as the robot will not move until we ask it to:

```
async def main():
  while not robot.check_imu_status():
    await asyncio.sleep(0.1)
  robot.uart.write("Ready to go!\n".encode())
  await control_loop()
asyncio.run(main())
```

Send this to the robot and calibrate it. Then, when you see **Ready to go** on Bluetooth, you can send back an angle to turn.

The same troubleshooting steps apply as before.

Try 30, 45, 60, and 90, or small values such as 5 and 10 degrees. Do not go above 179 or -179, as this can cause the robot to spin until turned off. You could add code to limit this.

It can be helpful to store the error in `IMUTurnController` (as `self.error`) and plot this data instead of the angle for tuning the PID.

You can now make a known turn.

## Summary

In this chapter, we investigated the IMU and how we can use it to control the heading of our robot. We learned how to connect the device and calibrate it.

Then, we used data from it to face North by combining the sensor data with a PID controller. Finally, we built on this example so that it can turn a specified number from the current heading.

In the next chapter, we will build a small arena for the robot and look at how we can combine the encoders and distance sensors to estimate the robot's position within this arena, improving its estimation as it moves.

## Exercises

These exercises will deepen your understanding of the topics that were covered in this chapter and make the robot code better:

- Combining the preceding behaviors with the menu system for the UART we've seen in previous chapters would allow you to tune the PID with the robot running.

- Could you use the known turn behavior and straight-line behavior to write a better version of the planned path program from *Chapter 5, Driving Motors with Raspberry Pi Pico*?

- Experiment with the Euler heading reading – after calibrating, see how the readings change when you bring the robot near objects such as a laptop or kitchen appliances. This experiment will demonstrate a weakness with this kind of sensor.

- An advanced experiment would be to extract the quaternion (instead of Euler data) and write this to the UART.

## Further reading

These further study aids will help you learn more and dive deeper into the PID algorithm and its quirks:

- The Adafruit CircuitPythong API guide for the BNO055 shows what else you can do with this sensor: `https://docs.circuitpython.org/projects/bno055/en/latest/api.html` - BNO055.

- *Learn Robotics Programming* provides a guide for interfacing a Raspberry Pi device with a different IMU chip, the ICM90248, and writing code to calculate Euler angles, along with interesting ways to visualize this. It also shows how you can use encoders to make a known turn instead, perhaps when objects distort the magnetometer readings.

- Paul McWhorter performs Arduino experiments with the same BNO055 sensor in an intensive video series: `https://toptechboy.com/arduino-based-9-axis-inertial-measurement-unit-imu-based-on-bno055-sensor/`.

# 13

# Determining Position Using Monte Carlo Localization

We now have several interesting sensors on our robot. However, we have yet to combine them to understand the position of our robot. The **Monte Carlo simulation** is a method that uses multiple sensors and a model of a robot's world to estimate its location and heading in that world.

You will learn how to make a test arena for a robot, followed by how to model this arena in code, and how to send this data over Bluetooth to view on a computer. You will practice statistical methods for the robot to start guessing its location. You will see how to enrich encoder data and move the guesses, and then integrate this with distance sensor data to refine the guesses, using a method that is effective in the face of noisy sensor data and can cope with minor inaccuracies. This will come together in a Monte Carlo guess and check loop.

In this chapter, we will cover the following main topics:

- Creating a training area for our robot
- Modeling a space
- Using sensors to track a relative pose
- Monte Carlo localization

## Technical requirements

For this chapter, you will require the following:

- The robot and code from *Chapter 12, Detecting Orientation with an IMU on Raspberry Pi Pico*
- A PC or laptop with Bluetooth LE
- Python 3.7 with the Python `matplotlib`, `bleak`, and `NumPy` libraries installed
- 10 x 10-mm A1 sheet foam boards

- Duct or gaffer tape
- A tape measure
- A metal ruler, set square, and pencil
- A sharp craft knife
- A floor space of 1.5 sq meters

You can find the code for this chapter at `https://github.com/PacktPublishing/Robotics-at-Home-with-Raspberry-Pi-Pico/tree/main/ch-13`.

# Creating a training area for our robot

We will be estimating a robot's location in a space. The robot needs a known space to work in, so we will build a simple world for it to operate in. This training area, or arena, is loosely based on those used in Pi Wars (see `https://piwars.org/2022-competition/general-rules/` under *Arena construction rules*), a British robotics competition, where this algorithm could be used for a robot to compete autonomously.

Let's take a closer look at the arena.

## What we will make

The following diagram shows the arena we will make:

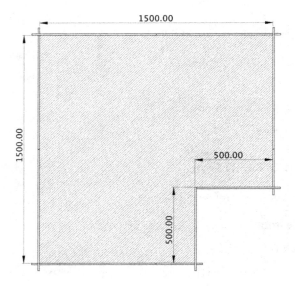

Figure 13.1 – A robot test arena

*Figure 13.1* shows a top-view drawing of an arena, complete with dimensions. The arena is mostly square to keep it simple to make and model. To help the Monte Carlo simulation work, there must be a cutout on one side to prevent rotational symmetry – that is, you can't rotate the arena and have it appear identical from multiple angles.

The arena should be large enough for the robot to move freely inside of it, without being excessively large, making 1,500 mm a good compromise. The arena walls should be tall enough that the robot's distance sensors cannot miss them. A reasonable wall height would be 200 mm. We will work with *mm* throughout this chapter to keep things consistent.

---

**Arena size versus robot speed**

Beware that you may want a larger arena for a faster robot, and that a smaller arena will give the robot less time to detect its features.

The arena floor surface is important; if the robot's wheels are slipping, then the calculations will suffer in accuracy.

---

Next, we can see how we'll build this.

## How we will make the arena

We'll use foam board to build the arena, as it is lightweight and easy to cut; A1 boards are readily available, and panels can be cut from these.

The following figure shows how we can make the arena:

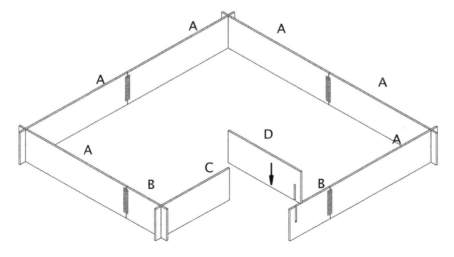

Figure 13.2 – Making an arena

*Figure 13.2* shows a 3D view of the arena. The letters indicate different parts. The Dumont Cybernetics team inspired this style. We can make the arena in sections, slotting together corner joints, as shown between panels *D* and *B*, or use tape (such as duct tape) to make hinging sections, such as those between panels *A* and *B*. This arena is 1,500 mm, so it can be disassembled and folded small when not in use.

The following diagram shows the parts we will need to make this:

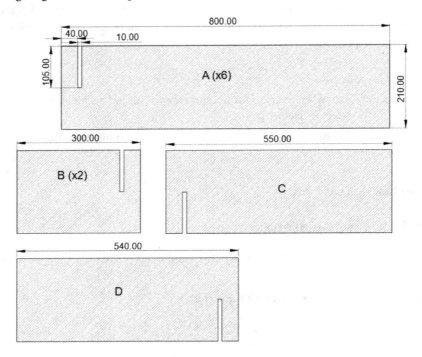

Figure 13.3 – A drawing of the parts to make the arena

*Figure 13.3* shows the parts to cut to make the arena. Each part has a letter, the number of pieces you'll need to make, and the measurements to cut the part. The slot profiles are all the same as panel **A**, along with the wall heights.

Four A panels can be cut from a board with some material left over. Let's see how to cut them.

## Tips for cutting

You can cut the foam board in a similar way to the plastics you cut in *Chapter 4, Building a Robot around Pico*. Use the tape measure, large ruler, and set square to mark where you will cut in pencil. Ensure the surface you are using to cut is at a comfortable height so that the long, repeated cutting does not make your back sore.

Then, following a straight metal edge, draw a sharp knife along the cut multiple times. For the first cut, aim only to score the top plastic layer, and then keep making cuts until you are through. Take care to cut the same area – this is a matter of letting later cuts follow the earlier cuts by holding the blade lightly over them.

I suggest cutting wall height strips first, before marking slots and wall lengths on them.

These sheets often come sandwiched with throw-away foam padding; this will help as a cutting surface so that you do not damage a table or floor underneath.

If there is tearing, either you are applying too much pressure or need to change your blade for a sharp, fresh one.

Take care cutting the slots. The wall heights do not need to be super precise; within a few mm is good enough. The real world is often not as precise and clear as a simulation, and this algorithm will be able to cope with this.

Once you have cut the parts, assemble the corners of the arena, and then make tape hinges on the inside joins (not the slots). When you disassemble the first time, fold the parts along these hinges, and then put tape on the outside of this joint. With gaffer tape or duct tape, this should be sturdy enough.

Now that we have a real space, we'll need to model this so that the robot can use it.

## Modeling the space

The aim of a Monte Carlo system is to model or simulate a space and a robot's location. In this section, we will learn how code for the robot will represent this space. We will also look at how a computer can be used to visualize our robot's guesses. Monte Carlo-based behavior code checks sensor readings frequently against the model of the space, so we should represent the space on the robot to optimize this.

The role of the computer and the robot in this are shown in the following diagram:

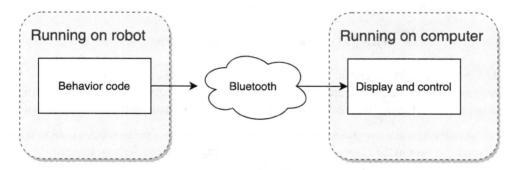

Figure 13.4 – Visualizing with the computer

*Figure 13.4* shows an overview of this system's display and control architecture. The behavior code runs on the robot. The computer displays the state of the robot code, along with start and stop controls. The arena and state of the system all belong to the robot.

Let's look at how to represent the arena on the robot.

## Representing the arena and robot position as numbers

For a model like this, the boundaries of the arena are important. We can start by taking 2D *X* and *Y* coordinates of the corners.

Look at the following representation of the arena:

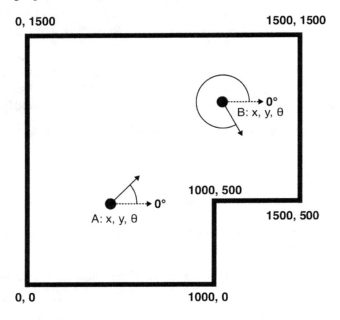

Figure 13.5 – The arena and poses as coordinates

*Figure 13.5* shows a simplified version of the arena. Coordinates describe each corner as numbers. We can directly use those in the code. These corners can be joined as line segments. A line segment is a set of coordinates for the start and end of the line segment. All the coordinates in the code will be in mm.

Our robot will have a **pose** somewhere within the arena. A pose describes the robot's location in space – in this case, anywhere in the 2D space of the arena and facing any of 360 degrees. Imagine this like a map pin, with an arrow point showing the heading.

*Figure 13.5* also shows two robot poses, *A* and *B*. Each has an *X* and *Y* coordinate in mm within the arena, and each has a heading theta (θ) in degrees. These three numbers will represent every robot pose in this 2D space. At the start of a simulation, the robot could be at any position and facing any heading within the arena in degrees.

Our arena representation has *0, 0* for the bottom left. Heading *0* faces right, with positive theta angles going anticlockwise. For example, pose *A* has a heading of around 30 degrees anticlockwise from the right, and pose *B* has a heading of 300 degrees from the right.

In this system, we will have many pose estimates, which behave like particles. The Monte Carlo simulation here is also known as a **particle filter**, due to how poses are manipulated and then filtered away based on sensor data.

Now, let's develop code for the arena boundary line segments.

### Converting the representation into code

We'll represent the arena as code and render it on the computer. Then, we'll move the arena representation over to the robot, with the computer fetching data from it.

After creating the `arena.py` file, we can add the arena points to it:

```
width = 1500
height = 1500
cutout_width = 500
cutout_height = 500

boundary_lines = [
    [(0,0), (0, height)],
    [(0, height), (width, height)],
    [(width, height), (width, cutout_height)],
    [(width, cutout_height), (width - cutout_width, cutout_
height)],
    [(width - cutout_width, cutout_height), (width - cutout_
width, 0)],
    [(width - cutout_width, 0), (0, 0)],
]
```

The `boundary_lines` variable represents a list of line segments, each of which is an array of start and end coordinates, read as `[(start_x, start_y), (end_x, end_y)]`. We also store the arena `width` and `height` values here. If your arena is a different size, please update these values.

We can display this using `matplotlib`, a mathematical plotting library for Python. To do this, first install Python 3.7 (or later) on your computer, and in a terminal, use the `python3 -mpip install matplotlib numpy` command to get the libraries. For Linux, you may need additional `python3-tk` packages in your package manager.

Create the `display_arena.py` file to draw the arena. This file starts by importing `matplotlib`. The convention is to import `pyplot`, a data-plotting module, as `plt`:

```
from matplotlib import pyplot as plt
import arena
for line in arena.boundary_lines:
    plt.plot([line[0][0], line[1][0]], [line[0][1], line[1]
[1]], color="black")
plt.show()
```

We will loop over the lines in the arena. The `plot` method takes $X$ coordinates for a line, followed by $Y$ coordinates for it, and allows us to specify a line color. Run this with `python3 display_arena.py`, which will draw the arena for us:

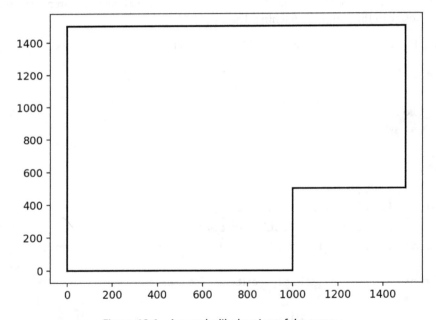

Figure 13.6 – A matplotlib drawing of the arena

The preceding diagram shows the arena drawn by the computer from our code. It has grid coordinates along the left and the bottom.

We can now look at moving this model data to the robot where it will be used.

## Serving the arena from the robot

The robot should be the source of truth for arena data, so let's put the arena model there. Make a `robot` folder on your computer and move `arena.py` into it. We will be copying the contents of this `robot` folder to Raspberry Pi Pico. From the previous chapters, copy `robot.py`, `pid_controller.py`, and `pio_encoder.py` into the `robot` folder.

We can then add a little code to serve up our arena boundary lines from the robot. In `robot/code.py`, start with imports and helpers:

```python
import asyncio
import json

import arena
import robot


def send_json(data):
    robot.uart.write((json.dumps(data)+"\n").encode())


def read_json():
    data = robot.uart.readline()
    decoded = data.decode()
    return json.loads(decoded)
```

On the robot, we can handle commands as we have been since *Chapter 10, Using the PID Algorithm to Follow Walls*; however, we will use **JavaScript Object Notation** (**JSON**), a convenient method to represent more complex information.

Any data we send is converted into JSON, and then a `"\n"` newline is added to show that it's a complete message. We then encode this. The data we receive is unpacked with `json.loads`, which will result in data structures of dictionaries.

We can then add a handler to this that will send back the arena when requested:

```python
async def command_handler():
    print("Starting handler")
    while True:
        if robot.uart.in_waiting:
            request = read_json()
            print("Received: ", request)
            if request["command"] == "arena":
```

```
        send_json({
            "arena": arena.boundary_lines,
        })
    await asyncio.sleep(0.1)
asyncio.run(command_handler())
```

This will loop and wait for the arena command. It prints out any JSON it receives for troubleshooting. It will use the JSON to send back the arena data.

The robot can be interacted with via the Bluefruit app in UART mode by sending `{ "command" : "arena" }`. The robot will send the boundary lines back as lists. However, ideally, we want the computer to display this from the robot with `matplotlib`. We'll need to connect the computer to the robot first.

## The Bleak library

The **Bleak** Python library allows Python on a computer to connect and interact with Bluetooth LE devices. In a terminal, use `python3 -mpip install bleak` to install this. Bleak is documented at `https://bleak.readthedocs.io/en/latest/`.

We will also need information about the Adafruit Bluefruit system. Bluetooth LE has device IDs, and IDs for services on Bluetooth. See `https://learn.adafruit.com/introducing-adafruit-ble-bluetooth-low-energy-friend/uart-service` for details. We will be using these in the following piece of code.

We'll start with an example to list the devices, to check whether we can find the robot's Bluetooth UART. Create the `find_devices.py` file and add the following:

```
import asyncio
import bleak


async def run():
    ble_uuid = "6E400001-B5A3-F393-E0A9-E50E24DCCA9E"
    ble_name = "Adafruit Bluefruit LE"
    devices = await bleak.BleakScanner.discover(service_
uuids=[ble_uuid])
    print(f"Found {len(devices)} devices")
    print([device.name for device in devices])
    matching_devices = [device for device in devices if device.
name==ble_name]
    if len(matching_devices) == 0:
```

```
        raise RuntimeError("Could not find robot")
    ble_device_info = matching_devices[0]
    print(f"Found robot {ble_device_info.name}...")
asyncio.run(run())
```

This code starts by importing the `asyncio` and `bleak` libraries. The run function needs to be asynchronous so that it can await the `bleak` scanner.

We define the ID and name of the Adafruit Bluefruit based on the Adafruit documentation, and then ask the `bleak` library to discover available devices with the Adafruit UART service. After waiting for the result, the next few lines print these out. The function then filters for the device with the matching name, checks that it found it, and prints it successfully.

Run this with `python3 find_devices.py`. If the robot is off, you will see a `Could not find robot` error. However, running with the robot turned on should show the following output:

```
Found 1 devices
['Adafruit Bluefruit LE']
Found robot Adafruit Bluefruit LE...
```

From time to time, `bleak` will have trouble finding the robot and display the preceding error. You will need to rerun the example to find the robot. We can now put this code into a library that we can use in the remaining experiments in this chapter.

## Creating a Bluetooth LE wrapper library

We'll call the library `robot_ble_connection.py`. We'll start with imports:

```
import asyncio
import bleak
```

We'll put our connection handling into a class:

```
class BleConnection:
    ble_uuid = "6E400001-B5A3-F393-E0A9-E50E24DCCA9E"
    rx_gatt = "6E400003-B5A3-F393-E0A9-E50E24DCCA9E"
    tx_gatt = "6E400002-B5A3-F393-E0A9-E50E24DCCA9E"
    ble_name = "Adafruit Bluefruit LE"
```

See https://learn.adafruit.com/introducing-adafruit-ble-bluetooth-low-energy-friend/gatt-service-details for an explanation of these variables.

When we create the object to handle the connection, we will have two functions that the client code can provide, one for a connection being complete and one for data being received:

```python
def __init__(self, receive_handler):
    self.ble_client = None
    self.receive_handler = receive_handler
```

`receive_handler` is a function that can be called with a Python `bytes` object holding the received data. We'll adapt our receive handler into one that the `bleak` library can use to receive data:

```python
def _uart_handler(self, _, data: bytes):
    self.receive_handler(data)
```

Now, we add a connect method. This starts the same as the `find_devices` example:

```python
async def connect(self):
    print("Scanning for devices...")
    devices = await bleak.BleakScanner.discover(
        service_uuids=[self.ble_uuid]
    )
    matching_devices = [device for device in devices if
device.name==self.ble_name]
    if len(matching_devices) == 0:
        raise RuntimeError("Could not find robot")
    ble_device_info = matching_devices[0]
    print(f"Found robot {ble_device_info.name}...")
```

However, we then need to connect to this device and handle the received data:

```python
    self.ble_client = bleak.BleakClient(ble_device_info.
address)
    await self.ble_client.connect()
    print("Connected to {}".format(ble_device_info.name))
    self.notify_task = asyncio.create_task(
        self.ble_client.start_notify(self.rx_gatt, self._
uart_handler)
    )
```

We create a `BleakClient` object and then wait for a connection to the robot. After connection, it will create a background task to notify the handler when data arrives. This `start_notify` method uses `rx_gatt` to receive UART data from this Adafruit Bluefruit device.

We need to be able to close the connection:

```
async def close(self):
    await self.ble_client.disconnect()
```

Then, the final part of this code can send data to the robot:

```
async def send_uart_data(self, data):
    await self.ble_client.write_gatt_char(self. tx_gatt,
data)
```

This will wait for data to be sent and use the right UUID for transmitting to the UART.

This robot_ble_connection library is now ready to be used in code.

## Showing the robot's data on the computer screen

We can use matplotlib to display the data from the robot, connecting to the robot with the preceding code, and asking it for the arena. This demonstration will tie matplotlib together with a Bluetooth connection.

We'll put this in a new file named display_from_robot.py, starting with the imports:

```
import asyncio
import json
import matplotlib.pyplot as plt
from robot_ble_connection import BleConnection
```

We'll put our display system in a class called RobotDisplay:

```
class RobotDisplay:
    def __init__(self):
        self.ble_connection = BleConnection(self.handle_data)
        self.buffer = ""
        self.arena = {}
        self.closed = False
        self.fig, self.axes = plt.subplots()
```

The first part sets up the BLE connection and prepares it with a handle_data method (this is the BLE data handler, which we'll implement shortly).

When data arrives via BLE to the computer, a whole message can be split across a few calls to the handle_data method. We are working in lines of text, so we will use self.buffer to store any partial line until we get a line ending, signaling a line is complete. We also have a place to store the arena from the robot, and a flag to detect when the app is closed. The display system is prepared with `plt.subplots`, which gets a figure and axes – we'll use these in a `draw` method to draw the display.

Let's make a handler for the app being closed:

```python
def handle_close(self, _):
    self.closed = True
```

This handler will just set the `closed` flag to `True`, which we can check for later. `matplotlib` will automatically create an app window for us to display output.

Next, we will build the BLE data handler:

```python
def handle_data(self, data):
    self.buffer += data.decode()
    while "\n" in self.buffer:
        line, self.buffer = self.buffer.split("\n", 1)
        print(f"Received data: {line}")
        try:
            message = json.loads(line)
        except ValueError:
            print("Error parsing JSON")
            return
        if "arena" in message:
            self.arena = message
```

This collects decoded incoming data into the `self.buffer` variable. While that buffer has line endings, `"\n"`, it splits a single line off and decodes it as JSON.

We then check whether this JSON has arena data in it. If so, we store it in the `arena` data member.

Next, we put the arena line drawing into a method:

```python
def draw(self):
    self.axes.clear()
    if self.arena:
        for line in self.arena["arena"]:
            self.axes.plot(
                [line[0][0], line[1][0]], [line[0][1],
line[1][1]], color="black"
            )
```

This function clears the previous display using the `self.axes.clear()` function and then redraws the arena lines.

The app `main` method starts the connection and asks the robot for the arena:

```
async def main(self):
    plt.ion()
    await self.ble_connection.connect()
    try:
        request = json.dumps({"command": "arena"}).encode()
        print(f"Sending request for arena: {request}")
        await self.ble_connection.send_uart_data(request)
        self.fig.canvas.mpl_connect("close_event", self.handle_close)
```

This function enables interactive mode in `matplotlib` with `plt.ion()` – this means we get to handle when the screen is redrawn, which suits our data model.

We then call and wait for the BLE `connect` function. Once a connection has been made, we wrap the rest in a `try`/`finally` block that will ensure the BLE connection is closed if this code is stopped or breaks. We then send a request to the robot, asking for the arena.

The code sets up the `close` handler so we can detect whether the window is closed, and immediately gets into a main `while` loop based on the closed flag:

```
        while not self.closed:
            self.draw()
            plt.draw()
            plt.pause(0.05)
            await asyncio.sleep(0.01)
    finally:
        await self.ble_connection.close()
```

The main loop uses `plt.draw()` to update the display and then waits 0.05 seconds, giving `matplotlib` time to handle interactive events. It also has a 0.01-second asynchronous sleep to give the BLE tasks time to run. These sleeps and pauses must be called frequently. At the end, `finally` ensures we close the BLE connection.

We then need to create an instance of the class and start the `main` loop:

```
robot_display = RobotDisplay()
asyncio.run(robot_display.main())
```

At this point, the display code is complete. Send the `robot` folder to Raspberry Pi Pico, and with battery power turned on, start the display code with the following:

```
python3 display_from_robot.py
```

You should see the BLE connecting messages and then the following output on the computer:

```
Sending request for arena: b'{"command": "arena"}'
Received data: {"arena": [[[0, 0], [0, 1500]], [[0, 1500],
[1500, 1500]], [[1500, 1500], [1500, 500]], [[1500, 500],
[1000, 500]], [[1000, 500], [1000, 0]], [[1000, 0], [0, 0]]]}
```

After around 30 seconds, you should see the computer display the arena. This will look identical to *Figure 13.6*, but the data is now coming from the robot.

We have the computer connecting to the robot and retrieving arena information from it. The robot has modeled the space in simple terms.

In the next section, we'll look more at robot poses, displaying them on our computer, and updating them from encoder sensors.

# Using sensors to track relative pose

In this section, we will explore what a pose is, how to create, send, and display poses, and how to move the poses relative to the movement of the robot.

## Setting up poses

We'll make some random poses in `robot/code.py` using **NumPy**, a numeric manipulation library for fast array operations, with `ulab` providing this functionality in CircuitPython. This library also gives us handy ways of storing and dealing with arrays.

Import the `ulab` library, and `random` to generate random poses:

```
import asyncio
import json
import random
from ulab import numpy as np
```

After the `read_json` function, we'll add a `Simulation` class to hold the poses:

```
class Simulation:
    def __init__(self):
        self.population_size = 20
```

```
        self.poses = np.array(
            [(
                int(random.uniform(0, arena.width)),
                int(random.uniform(0, arena.height)),
                int(random.uniform(0, 360))) for _ in
range(self.population_size)],
            dtype=np.float,
        )
```

We will create a small population of 20 random poses. The `poses` variable is a NumPy array of `population_size` items, with each item an *X*, *Y* heading pose. NumPy allows us to specify a datatype; we use the `float` type so that we can work in fractional values.

Add a function (before `Simulation`) to send *X* and *Y* pose coordinates to the computer:

```
def send_poses(samples):
    send_json({
        "poses": np.array(samples[:,:2], dtype=np.int16).
tolist(),
    })
```

The `[:, :2]` notation lets us extract the first two entries of each pose in the `poses` array, the *X* and *Y* coordinates. We convert this to `int16` to reduce how much data is being sent – the UART is easily overwhelmed by pose data.

The command handler can now send poses after the arena for now:

```
async def command_handler(simulation):
    print("Starting handler")
    while True:
        if robot.uart.in_waiting:
            request = read_json()
            print("Received: ", request)
            if request["command"] == "arena":
                send_json({
                    "arena": arena.boundary_lines,
                })
                send_poses(simulation.poses)
        await asyncio.sleep(0.1)
simulation = Simulation()
asyncio.run(command_handler(simulation))
```

Now, `command_handler` has the simulation passed into it and sends the poses back after the arena. Before we start the handler, we create `simulation` from its class.

This code is now ready for the computer to display these poses.

## Displaying poses

We can now enhance our `matplotlib` file, `display_from_robot.py`, with the poses. First, we will add numpy to the imports:

```
import asyncio
import json
import numpy as np
import matplotlib.pyplot as plt
```

When we set up the display in the `__init__` method, we add an empty `poses` member:

```
        self.fig, self.axes = plt.subplots()
        self.poses = None
```

Next, we need to extend `handle_data` to load poses into an `int16` NumPy array:

```
        if "arena" in message:
            self.arena = message
        if "poses" in message:
            self.poses = np.array(message["poses"],
dtype=np.int16)
```

We then add extend the `draw` method to display poses, checking whether any are loaded and, if so, putting them into a scatter plot, slicing into the *X* and *Y* components to fit `matplotlib`:

```
        if self.arena:
            for line in self.arena["arena"]:
                self.axes.plot(
                    [line[0][0], line[1][0]], [line[0][1],
line[1][1]], color="black"
                )
        if self.poses is not None:
            self.axes.scatter(self.poses[:,0], self.poses[:,1],
color="blue")
```

Send the `robot` folder over to Pico, and then run `display_from_robot.py` on the computer, and after the BLE startup, you should see something like the following screenshot:

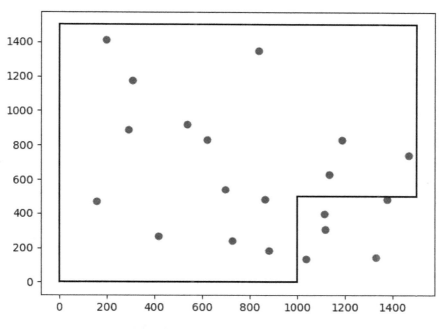

Figure 13.7 – Poses drawn in the arena

*Figure 13.7* shows the arena with 20 poses drawn as dots. Each dot is a potential guess of where the robot might be. Some are in the cutout area and will later be eliminated.

These poses will need to move when our robot moves, so let's make our robot move.

## Moving with collision avoidance

The robot will be moving while we perform the simulation, and it would be good to avoid collisions while the robot moves. We'll do this as an asynchronous routine so that other parts of the code can run at the same time. The following architecture diagram shows how this works:

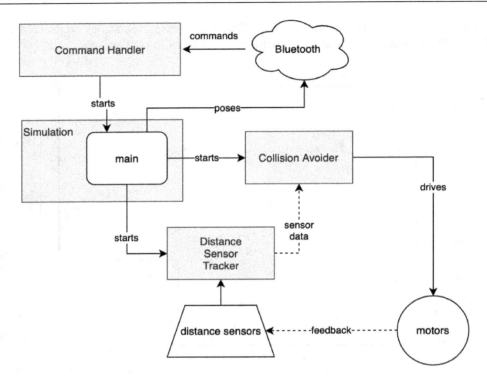

Figure 13.8 – A simulation with a collision avoidance architecture

The `command handler` system accepts Bluetooth command requests. The command handler starts the `main` loop in `simulation`. The simulation will start both a **Collision Avoider** and a **Distance Sensor Tracker**, as shown in *Figure 13.8*. The distance sensor tracker will store sensor data used by the simulation and the collision avoider. The collision avoider will drive the robot's motors. The `simulation main` method also sends poses via Bluetooth.

We will start with `DistanceSensorTracker`, a class to keep tabs on the distance sensors and their last readings. Place this in `robot/code.py` under the imports:

```
class DistanceSensorTracker:
    def __init__(self):
        robot.left_distance.distance_mode = 2
        robot.right_distance.distance_mode = 2
        self.left = 300
        self.right = 300
```

We are being explicit about sensor mode here, adjusting it to the size of the arena. We also put in starting values until a reading is available.

The sensor tracker loop fetches readings when ready and resets the sensor interrupts:

```
async def main(self):
    robot.left_distance.start_ranging()
    robot.right_distance.start_ranging()
    while True:
        if robot.left_distance.data_ready and robot.left_
distance.distance:
            self.left = robot.left_distance.distance * 10
            robot.left_distance.clear_interrupt()
        if robot.right_distance.data_ready and robot.right_
distance.distance:
            self.right = robot.right_distance.distance * 10
            robot.right_distance.clear_interrupt()
        await asyncio.sleep(0.01)
```

We are multiplying the sensor readings by 10 to convert them to mm, and then storing them. The remaining code can just use these stored readings.

Next, we'll build the `CollisionAvoid` class to turn the robot away from a wall that it detects with the sensors. Add this class after the `DistanceSensorTracker` class:

```
class CollisionAvoid:
    def __init__(self, distance_sensors):
        self.speed = 0.6
        self.distance_sensors = distance_sensors
```

This has an initial robot speed, along with a reference to the distance sensor tracker. This then has a `main` collision-avoiding loop:

```
async def main(self):
    while True:
        robot.set_right(self.speed)
        while self.distance_sensors.left < 300 or \
                self.distance_sensors.right < 300:
            robot.set_left(-self.speed)
            await asyncio.sleep(0.3)
        robot.set_left(self.speed)
        await asyncio.sleep(0)
```

This loop starts the right motor moving, and if a collision looks likely, it will set the left motor going backward and wait before driving forward. The `asyncio.sleep` delays mean that other tasks can continue on the robot.

Inside the `Simulation` class, add the sensors and `collision_avoider` to the `__init__` method:

```
        self.distance_sensors = DistanceSensorTracker()
        self.collision_avoider = CollisionAvoid(self.distance_
sensors)
```

Then, we add a simulation `main` method below the `__init__` simulation. This starts tasks for the other components and then loops over, sending the poses back to the computer:

```
    async def main(self):
        asyncio.create_task(self.distance_sensors.main())
        collision_avoider = asyncio.create_task(self.collision_
avoider.main())
        try:
            while True:
                await asyncio.sleep(0.1)
                send_poses(self.poses)
        finally:
            collision_avoider.cancel()
            robot.stop()
```

There's also error handling to stop the robot if anything goes wrong here – we cancel the collision avoider task (which would set the robot's speed) and stop the motors. The sleep here allows the other tasks to run and avoids overwhelming the BLE UART.

Extend the `command_handler` method to start the simulation's `main` task. We'll do so based on a **Start** button in the display UI. First, we'll store the task state at the top of the handler:

```
async def command_handler(simulation):
    print("Starting handler")
    simulation_task = None
    while True:
```

Then, we'll handle a `start` command in it:

```
        if request["command"] == "arena":
            send_json({
                "arena": arena.boundary_lines,
```

```
                  })
        elif request["command"] == "start":
            if not simulation_task:
                simulation_task = asyncio.create_
task(simulation.main())
```

The start button will run the simulation `main` task if it's not yet been run.

### Adding the start button on the computer

We need to add the corresponding button to the computer display code. Open `display_from_robot.py`. In the imports, add the following:

```
from matplotlib.widgets import Button
```

In the `RobotDisplay` class, we can add a helper to send a JSON command, much as we did on the robot. Add this to the robot display class above its `main` method:

```
async def send_command(self, command):
    request = (json.dumps({"command": command}) ).encode()
    print(f"Sending request: {request}")
    await self.ble_connection.send_uart_data(request)
```

This must be asynchronous to use `await` on the BLE `send_uart_data` function.

Above `main`, add a start button handler to call when the button is pressed:

```
def start(self, _):
    self.button_task = asyncio.create_task(self.send_
command("start"))
```

This will start sending the data but not wait for it – so the `matplotlib` event loop doesn't get stuck waiting.

We can replace the JSON sending in the `main` method with the `send_command` method:

```
await self.ble_connection.connect()
try:
    await self.send_command("arena")
```

We then add the button. Add the highlighted code into the `main` method:

```
        self.fig.canvas.mpl_connect("close_event", self.
handle_close)
        start_button = Button(plt.axes([0.7, 0.05, 0.1,
0.075]), "Start")
        start_button.on_clicked(self.start)
        while not self.closed:
```

The code uses the `send_command` wrapper to request the arena on startup. We then add `start_button`, using `plt.axes` to position it.

We connect a button `on_clicked` handler to the `start` method to enable the button.

Send the `robot` folder to Raspberry Pi Pico, and on the computer, run `display_from_robot.py`. I recommend propping the robot up for troubleshooting while connected, and then test it in the arena. The display will look like the following screenshot:

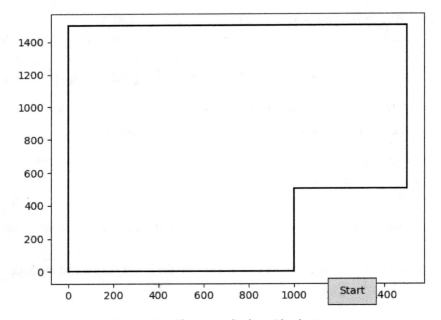

Figure 13.9 – The arena display with a button

*Figure 13.9* shows the **Start** button on the display. Press **Start** to get the robot running – poses will appear, and the robot should be avoiding walls in the arena. If not, the next section will help.

### Troubleshooting

These suggestions will help if you are having trouble here:

- If the distance sensors are showing errors, please go back to *Chapter 8, Sensing Distances to Detect Objects with Pico*, and check the wiring, using the tests there.

- If the robot is turning too far and getting trapped in the corners, lower the sleep after `robot.set_left(-self.speed)`.

- If the robot is going very quickly, either reduce `self.speed` or replace the motors with a greater gear ratio – ideally, 298:1, as recommended in *Chapter 11* in the *Slowing the Robot Down* section.

While the robot is now avoiding walls, the poses are not changing when the robot moves. To remedy this, we can add a motion model in the next section.

## Moving poses with the encoders

We want the poses to move with our robot's motion, updating both their position and their heading. The wheel encoders provide data about each wheel's motion, and we can convert this into rotations and translations of the pose. First, we need to store more data about the shape of the chassis in `robot/robot.py`:

```
ticks_per_revolution = encoder_poles * gear_ratio
ticks_to_mm = wheel_circumference_mm / ticks_per_revolution
ticks_to_m = ticks_to_mm / 1000
m_to_ticks = 1 / ticks_to_m
wheelbase_mm = 170
```

We ensure our tick conversion is in mm. We then add the wheelbase – this is a measurement between the central contact point of each wheel. Use a value measured from your own robot. We can use the wheelbase to calculate the robot's movement from the encoders, as shown in the following diagram:

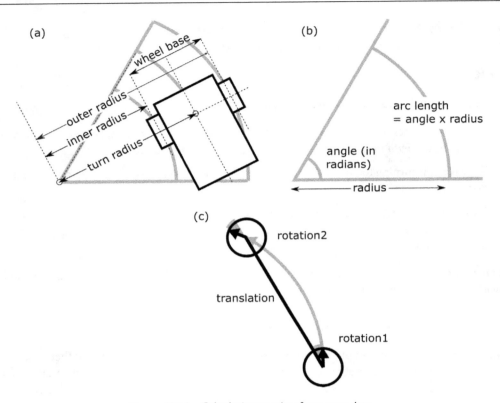

Figure 13.10 – Calculating motion from encoders

*Figure 13.10 (a)* shows the robot moving along an arc. Each wheel encoder will sense travel along an arc on an inner radius or an outer radius. Our robot uses a differential drive, so we can assume all motion takes place around the axis between the two wheels. The center of the robot, our pose, travels along the turn radius. We can use these with the wheel distance – the distance between the two wheels to calculate the arc.

*Figure 13.10 (b)* relates an arc length to the arc angle and radius. Each wheel will have traveled an arc, and the encoder will have measured the arc length. This arc length is the radius multiplied by the angle (in radians). We will use this to calculate the arc. From the motion of the two wheels, measured by the encoders (arc lengths) and the wheel distance, we can get the radius and angle change (d_theta).

*Figure 13.10 (c)* represents a robot motion. Although the robot has moved in an arc, for the simulation, we will simplify this arc motion into three components – *rotation 1* aligns the robot for a straight-line *translation*, and then *rotation 2* turns the robot to face the heading expected at the end of the curve.

Add the following into the `Simulation` class in `robot/code.py`:

```
def convert_odometry_to_motion(self, left_encoder_delta,
right_encoder_delta):
    left_mm = left_encoder_delta * robot.ticks_to_mm
    right_mm = right_encoder_delta * robot.ticks_to_mm
    if left_mm == right_mm:
        return 0, left_mm, 0
```

This function will take the changes (or deltas) in encoders and convert them to obtain representations of *rotation 1*, *translation*, and *rotation 2*. The encoder changes are turned into measurements in mm. We then check for the straight-line case, and if it is there, return a translation component only. This prevents 0 causing the next part to crash.

The remaining cases now need us to calculate an arc:

```
radius = (robot.wheelbase_mm / 2) * (left_mm + right_
mm) / (right_mm - left_mm)
    d_theta = (right_mm - left_mm) / robot.wheelbase_mm
    arc_length = d_theta * radius
```

The first line uses the wheelbase and the two encoder movements to calculate an arc radius. The difference between the motion of the two wheels is then used to calculate `d_theta`, how much the robot's heading changed throughout this arc. The `d` prefix represents a delta.

The arc length is then `d_theta` multiplied by the radius. Because this will be called fairly frequently, we are going to assume that the arc length is close enough to the translation.

From here, the rotation components can be calculated:

```
rot1 = np.degrees(d_theta/2)
rot2 = rot1
return rot1, arc_length, rot2
```

If we assume the arc to be regular, then each rotation component is half of the full arc rotation. We also convert this into degrees.

We can then write a method to move poses this way. Add this to `Simulation` by following these steps:

1.  Define the method and add the first rotation into the array of pose rotations (the third element):

```
def move_poses(self, rot1, trans, rot2):
    self.poses[:,2] += rot1
```

2. We then move the translation term in the new pose direction:

```
rot1_radians = np.radians(self.poses[:,2])
self.poses[:,0] += trans * np.cos(rot1_radians)
self.poses[:,1] += trans * np.sin(rot1_radians)
```

The `rot1_radians` variable will hold a NumPy array. This comes from the second element of the poses array, converted into radians. The ability of NumPy (or ulab) to operate on whole arrays is handy here. We will use it again to calculate the *X* and *Y* motions. `trans * np.cos` applies the cosine function to every element in `rot1_radians` and multiplies each one by the translation term.

3. We then need to add the `rot2` term:

```
self.poses[:,2] += rot2
self.poses[:,2] = np.array([float(theta % 360)
for theta in self.poses[:,2]])
```

Finally, we constrain the angles between 0 and 360 degrees.

4. Next, we need to tie these together with getting the encoder deltas. First, we extend the `Simulation.__init__` method to get the initial encoder readings:

```
self.collision_avoider = CollisionAvoid(self.
distance_sensors)
self.last_encoder_left = robot.left_encoder.
read()
self.last_encoder_right = robot.right_encoder.
read()
```

We will use this encoder data in a motion model, moving all our poses with the robot's motion. In the `Simulation` class, we will then add a `motion_model` method:

5. It first gets the latest encoder readings:

```
def motion_model(self):
    new_encoder_left = robot.left_encoder.read()
    new_encoder_right = robot.right_encoder.read()
```

6.  We calculate the deltas and feed these into `convert_odometry_to_motion`:

```
        rot1, trans, rot2 = self.convert_odometry_to_
   motion(
            new_encoder_left - self.last_encoder_left,
            new_encoder_right - self.last_encoder_right)
```

7.  We must update the last encoder readings so that we'll get deltas next time:

```
        self.last_encoder_left = new_encoder_left
        self.last_encoder_right = new_encoder_right
```

8.  Now, we apply these odometry values to our poses:

```
        self.move_poses(rot1, trans, rot2)
```

We now need to call this motion model. In `Simulation.main`, add the highlighted code:

```
        while True:
            await asyncio.sleep(0.05)
            self.motion_model()
            send_poses(self.poses)
```

This will apply the motion model at every cycle before sending the poses. Since the motion model requires time to run, the sleep is reduced to compensate, keeping the UART data rate similar. Copy the contents of the `robot` folder to Raspberry Pi Pico, and launch the `display_from_robot` app.

When you press **Start**, you should now see the poses moving as the robot moves around the arena. All the poses should follow the same path, but each from a different starting point and orientation.

These poses are moving, but the real world is messy, so let's add randomness to this.

## Pose movement probabilities

Robot movement is not always certain; although we have encoders, wheels can slip, wheel sizes can have minor variations, and our calculations might not be 100% accurate. Imagine that the preceding poses are in a cluster or the cloud, and then our robot drives in a particular direction. The following diagram should demonstrate this:

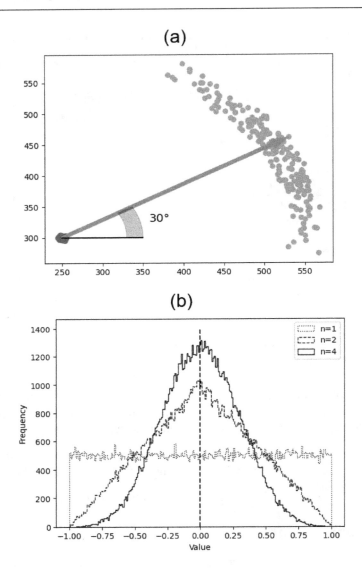

Figure 13.11 – Movement probability distributions

*Figure 13.11 (a)* shows a section of the arena, with arena coordinates shown on the axes. The cluster around the point (250,300) is the initial robot pose guesses. The thick line and angle arc show a robot movement at 300 mm, bearing 30 degrees. However, due to the uncertainties of the motion, the cluster gets spread out. The arc shape is due to uncertainty in the angle of the motion, and the width of the arc represents the uncertainty of the forward motion of the robot. This banana shape represents where a robot could end up. The image here has been exaggerated, as the spread on the robot should be far less than this.

*Figure 13.11 (b)* shows how we can model this uncertainty centered around a **mean** (average or most likely value) of 0, with a variation on either side. A **probability distribution** maps how likely a value is to come up in a random selection. The height of each point signifies how likely a particular value is to come up. If we use a uniform distribution, all possibilities between -1.0 and +1.0 are equal, giving us a rectangle shown for *n=1*. However, we want this distribution centered around the mean. If we sum two samples from the uniform distribution from -1.0 to 1.0 and divide by 2, we get the *n=2* graph. This is approximately a triangle. Here, *n* represents the number of uniform random sample picks we add together. We could refine this to the *n=4* curve, using a sum of four uniform samples and dividing by 4; however, the trade-off between ideal curves and the time cost for each uniform distribution sample makes the triangle at *n=2* good enough for our purposes to center the distribution.

We will use the *n=2* distribution in our model. In `robot/code.py`, add the following piece of code before `class Simulation`:

```
def get_random_sample(mean, scale):
    return mean + (random.uniform(-scale, scale) + random.
uniform(-scale, scale)) / 2
```

This code will add the two samples, scaling the uniform distributions to match how much the model varies, divide by 2, and then add the mean.

The other factor we will need to account for is that the larger the movement we make, the larger the random error factor will be. A large forward motion will influence the rotation, and large rotations will affect the forward motion (translation). It is conventional to refer to the factors for these influences as `alpha`. Let's add these values to our `Simulation.__init__` class:

```
self.last_encoder_right = robot.right_encoder.read()
self.alpha_rot = 0.09
self.alpha_rot_trans = 0.05
self.alpha_trans = 0.12
self.alpha_trans_rot = 0.05
```

We have four terms here. They should be values between 0 and 1 and kept low. The value `0.05` will represent a 5% error. Tune them to reflect the error seen in your robot.

We will use this to apply randomness to our model:

1.  Add the following method in `Simulation`, after `move_poses`:

```
def randomise_motion(self, rot1, trans, rot2):
```

2.    Calculate the scaling factors from the `alpha` terms:

```
rot1_scale = self.alpha_rot * abs(rot1) + self.
alpha_rot_trans * abs(trans)
trans_scale = self.alpha_trans * abs(trans) +
self.alpha_trans_rot * (abs(rot1) + abs(rot2))
rot2_scale = self.alpha_rot * abs(rot2) + self.
alpha_rot_trans * abs(trans)
```

The rotation scaling factors are based on the absolute value of each element; they must not be negative. The rotation scales have both a rotation factor and a lower translation factor. The translation scale has a translation factor (usually larger) and a factor based on both rotations.

3.    We will now use this to generate noise around the motion for every pose:

```
rot1_model = np.array([get_random_sample(rot1,
rot1_scale) for _ in range(self.poses.shape[0])])
trans_model = np.array([get_random_sample(trans,
trans_scale) for _ in range(self.poses.shape[0])])
rot2_model = np.array([get_random_sample(rot2,
rot2_scale) for _ in range(self.poses.shape[0])])
```

This uses the scaled sample function we created, with our scale factor. It uses the calculated rotation or translation as a mean. We run this through loops for each pose dimension, so for a population of 200, we will get 200 random samples, centered around the calculated measurement, with variation scaled to the calculated factor.

4.    Finally, we return these models:

```
return rot1_model, trans_model, rot2_model
```

We now have a model that generates noise in our motion, meaning that it will compensate for the inaccuracies in the measurement by modeling the uncertainty in that measurement. Add the highlighted code for this to the `motion_model` method:

```
def motion_model(self):
    """Apply the motion model"""
    new_encoder_left = robot.left_encoder.read()
    new_encoder_right = robot.right_encoder.read()
    rot1, trans, rot2 = self.convert_odometry_to_motion(
        new_encoder_left - self.last_encoder_left,
        new_encoder_right - self.last_encoder_right)
    self.last_encoder_left = new_encoder_left
    self.last_encoder_right = new_encoder_right
```

```
        rot1_model, trans_model, rot2_model = self.randomise_
motion(rot1, trans, rot2)
        self.move_poses(rot1_model, trans_model, rot2_model)
```

The changes swap the `rot1` variable out for `rot1_model` and do a similar swap for the other pose elements. As `rot1_model` has the same number of elements as our poses, passing this into `move_poses` will add each sample element-wise to the respective pose element. This method takes advantage of how NumPy manipulates lists.

Copy the `robot` folder to the robot and run the `display_from_robot.py` app on your computer. The motion will be a little randomized. Now, let's check that your robot code is working and behaving as expected.

### Troubleshooting

If this example does not work, try the following methods:

- Run with the robot propped up and connected to the computer so that the Mu editor serial can see its state. This will show you whether there are any errors in the code.

- If the movements are too far or too little, adjust the measurements in `robot/robot.py` to match your robot, as they may vary from example values.

- If you see sensor or I2C issues, backtrack to check the wiring and previous sensor demonstration examples. Also, ensure that you have fresh batteries.

We now have our poses motion model based on the encoders. We can now bring the distance sensors and arena data into play with the Monte Carlo simulation.

# Monte Carlo localization

Our robot's poses are going outside of the arena, and the distance sensor readings should show which guesses (poses) are more likely than others. The **Monte Carlo** simulation can improve these guesses, based on the sensor-reading likelihood.

The simulation moves the poses and then observes the state of the sensors to create weights based on their likelihood, a process known as the **observation model**.

The simulation **resamples** the guesses by picking them, so those with higher weights are more likely. The result is a new generation of guesses. This movement of particles followed by filtering is why this is also known as a **particle filter**.

Let's start by giving our poses weights, based on being inside or outside the arena, and then we'll look at how to resample from this.

## Generating pose weights from a position

The initial weight generation can be based on a simple question – is the robot inside the arena or not? If not, then we can reduce the pose probability. Note that we don't eliminate these, as the robot could have been placed outside the arena map or been tested on your desk. We will just give them a lower probability than those that are inside the arena.

In the `robot/arena.py` file, do the following:

1.  We can add a value to indicate a very low probability – close to but not zero:

    ```
    low_probability = 10 ** -10
    ```

2.  Add a function to check whether the arena contains a point:

    ```
    def contains(x, y):
    ```

3.  First, check whether the point's outside the arena rectangle:

    ```
    if x < 0 or x > width \
      or y < 0 or y > height:
        return False
    ```

4.  Then, we check whether it's in the cutout section:

    ```
    if x > (width - cutout_width) and y < cutout_height:
        return False
    ```

5.  Otherwise, this point is in the arena:

    ```
    return True
    ```

We can then add an `observation_model` method to our `robot/code.py` `Simulation` class to generate the weights:

1.  We set up the weights to ones, with a weight per pose:

    ```
    def observation_model(self):
        weights = np.ones(self.poses.shape[0], dtype=np.
    float)
    ```

2.  We can then loop over the poses, lowering the weights of those outside the arena:

    ```
    for index, pose in enumerate(self.poses):
        if not arena.contains(pose[0], pose[1]):
            weights[index] = arena.low_probability
    ```

3.  We then return the weights:

```
return weights
```

At this point, the weights aren't being used. We will need to resample them for them to act on the poses.

## Resampling the poses

As we step through the Monte Carlo simulation on the robot, we would like a subsequent generation of particles to favor more likely poses. We are going to use a technique illustrated in the following diagram:

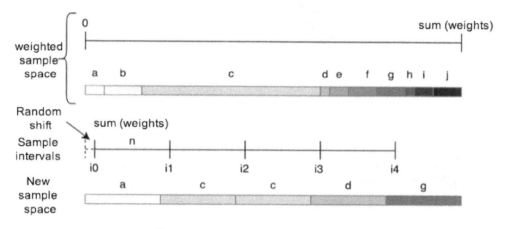

Figure 13.12 – Low variance resampling

*Figure 13.12* starts with **Weighted Sample Space**, a number line between 0 and the sum of all weights. Below this is a bar representing 10 samples (named *a* to *j*) in a sample space. The weights of these samples are represented by their widths. The shading highlights the different samples.

In the diagram, we generate a new space with five samples *(n=5)*. This number could be the same as the original space (for generating a new generation), may have a smaller number for sending via BLE, and may have a larger number for interpolating.

Resampling the original set starts by dividing the total sum by the number of new samples, which will give a sample interval size of *sum/n*, shown as **Sample intervals**. We then generate a single uniform random number between 0 and *sum/n*, which will shift the intervals.

We can then look at the weighted sample space and pick out the sample that matches the start of each interval – this is the **weight index**. This will produce **New sample space**. Note that sample *c*, which has the highest weight, gets sampled more times. With larger sample populations, the resampled space will more accurately resemble the original.

The new samples do not have weights and are all considered equally weighted, but some samples appear multiple times to represent their previous weight.

This technique of using random shifted intervals is known as the **low variance resampling** method. We will now see how to perform this through code:

1.  In `robot/code.py`, inside the `Simulation` class, add the following:

    ```
    def resample(self, weights, sample_count):
        samples = np.zeros((sample_count, 3))
    ```

    The `weights` variable refers to the list of weights, and `sample_count` refers to the number of samples to get. This method will sample new poses from the poses array. We will set up a `samples` variable to hold the new samples.

2.  Next, we set up the interval size based on `sample_count`:

    ```
    interval = np.sum(weights) / sample_count
    shift = random.uniform(0, interval)
    ```

    We can use that to set the interval `shift` value – the start position.

3.  We are going to store the cumulative weights while we loop through the original samples (poses). We will also store an index in the source sample set:

    ```
    cumulative_weights = weights[0]
    source_index = 0
    ```

4.  The code will loop until we have the expected number of samples. For each sample, there is a `weight_index` parameter:

    ```
    for current_index in range(sample_count):
        weight_index = shift + current_index * interval
    ```

5.  We now start adding up weights from the source samples in `cumulative_weights`, until they meet the weight index:

    ```
    while weight_index > cumulative_weights:
        source_index += 1
        source_index = min(len(weights) - 1, source_index)
        cumulative_weights += weights[source_index]
    ```

    We keep track of the source sample index that met this weight requirement.

6. We can use this `source_index` to add a sample to our set:

```
samples[current_index] = self.poses[source_
index]
```

This will drop out of the `while` loop and be the end of the `for` loop.

7. Finally, we return the new set of samples:

```
return samples
```

We can also increase our population, while sending only a subset. In `Simulation.__init__`, change the population size:

```
def __init__(self):
    self.population_size = 200
```

We are limiting the population size here due to Pico memory constraints.

We can then apply our observation model in the `main` loop (in `Simulation.main`):

```
while True:
    weights = self.observation_model()
    send_poses(self.resample(weights, 20))
    self.poses = self.resample(weights, self.
population_size)
    await asyncio.sleep(0.05)
    self.motion_model()
```

In our loop, we use the observation model to get weights for the poses. We use the `resample` method to get 20 poses to send. We then use `resample` again to get a new population of poses. The cycle of updating our poses, observing their state, weighting the poses, and then resampling them is known as a **recursive Bayes filter**.

If you send the `robot` folder to Raspberry Pi Pico and launch the app, you should start seeing the number of samples outside the arena being reduced. They will jump around, since we are sampling 20 from a larger set of 200.

The system reduces the sample space to those that are more likely. However, we can include distance sensors to improve this process.

## Incorporating distance sensors

Our robot has two distance sensors. Our code will check distances the robot senses against model data. If each pose is an imaginary map pin, then the distance to the nearest obstacle would be a string stretched out from this pin, or a sensor beam with a sensed endpoint – the **beam endpoint model**. With 200 poses, this could be slow. Let's see a faster method to model them.

### Modeling distance sensors in our space

One way we could do this is to take an estimate of the robot's position, and then perform the math needed to get the distance to the nearest obstacle. However, we can make a lookup table to simplify this.

Let's look at the following diagram:

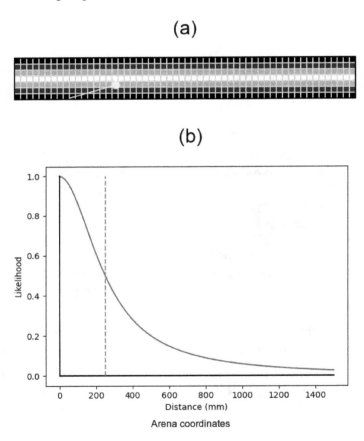

Figure 13.13 – Distance representation

*Figure 13.13 (a)* shows a **likelihood field** based on distance. The white dot is the endpoint at the distance reading from the distance sensor. The bright middle line of this grid represents an arena wall. Each grid square has a value between 0 and 1 (shown as brightness), representing how likely a sensed distance projected here is to have detected the wall. Instead of asking the question, *Is this distance measurement a match?*, we can ask, *How likely is this distance sensor a match?* We can calculate this grid once only when the system starts up, so other code can make fast lookups into the grid to check for sensor readings.

*Figure 13.13 (b)* shows how the likelihood changes, with a distance from 0 to 1,500 in a decaying function. The *Y* axis shows the likelihood of it being a hit. The dashed vertical line is a value, currently 250, at the inflection point, at which the curve changes direction. A smaller inflection point makes a tighter curve; a larger value makes a wider curve.

Let's import numpy at the top of robot/arena.py:

```
from ulab import numpy as np
```

We will convert these values to a grid of 50 mm² cells. As some poses will have distance endpoints outside the boundary, we'll give the grid an overscan. Extend the robot/arena.py library with the following:

```
grid_cell_size = 50
overscan = 2
```

We'll start with how we get the distance to a grid square. In robot/arena.py, after defining boundary_lines, add the following:

```
def get_distance_to_segment(x, y, segment):
    x1, y1 = segment[0]
    x2, y2 = segment[1]
    if y1 == y2 and x >= min(x1, x2) and x <= max(x1, x2):
        return abs(y - y1)
    if x1 == x2 and y >= min(y1, y2) and y <= max(y1, y2):
        return abs(x - x1)
    return np.sqrt(
        min(
            (x - x1) ** 2 + (y - y1) ** 2,
            (x - x2) ** 2 + (y - y2) ** 2
        )
    )
```

The code unpacks the line segment into `x1` and `y1`, and `x2` and `y2`. It then checks whether the line is horizontal (the same *Y*), and whether the point being checked is above or below it; this allows a shortcut by subtracting the *Y* values. The code can repeat this for vertical lines.

The code then uses Pythagoras' theorem, where the resulting distance will be the hypotenuse.

We will convert distances into likelihoods with a decaying function, which will return a lower value as we get further from zero. The following function will, for a specific point, find the nearest segment distance and then apply a decaying function to it:

```python
def get_distance_likelihood(x, y):
    min_distance = None
    for segment in boundary_lines:
        distance = get_distance_to_segment(x, y, segment)
        if min_distance is None or distance < min_distance:
            min_distance = distance
    return 1.0 / (1 + min_distance/100) ** 2
```

We can make a function to generate this grid. It starts by making a 2D array of float fields filled with zeros to hold the grid values:

```python
def make_distance_grid():
    grid = np.zeros((
            width // grid_cell_size + 2 * overscan,
            height // grid_cell_size + 2 * overscan
        ), dtype=np.float)
```

The width of the grid in cells is the width of the arena divided by the cell sizes. We then add in the overscan for either side of the arena. The height uses a similar calculation.

We then loop over the grid rows and columns to fill in the cell data:

```python
    for x in range(grid.shape[0]):
        column_x = x * grid_cell_size - (overscan * grid_cell_
size)
        for y in range(grid.shape[1]):
            row_y = y * grid_cell_size - (overscan * grid_cell_
size)
            grid[x, y] = get_distance_likelihood(
                column_x, row_y
            )
```

```
        return grid
distance_grid = make_distance_grid()
```

This loop gets the arena $X$ and $Y$ coordinates for each cell and uses the coordinates to fetch the likelihood at that position, storing it in the grid. We call this function and store the result in the `distance_grid` variable so that this calculation will run when the `arena.py` file is imported. The calculated distance grid looks like the following diagram:

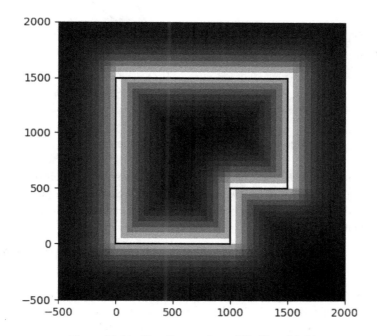

Figure 13.14 – The distance-based likelihood field

*Figure 13.14* shows the distance-based likelihood field.

The overscan extends to -500 mm and 2,000 mm, and the boundary lines are drawn in black. Each cell's value is based on its bottom-left coordinate.

We can now use this likelihood field in the observation model.

### Generating weights from the distance sensors

For each pose, we will need to project the sensed distances from the sensor positions. See the following diagram:

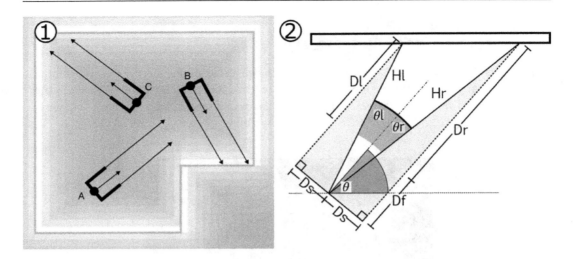

Figure 13.15 – Distance sensor geometry

*Figure 13.15 (1)* shows poses *A*, *B*, and *C* as dots, with an arrow showing their heading. There are dark bars representing the location of the two distance sensors relative to the pose – they stick out to the side and forward. From this location, we will have sensor readings, represented by the arrows pointing from the front of the bars.

By projecting the sensor distances against the likelihood field, we can see that pose B is a more likely match than pose C, with A as the least likely pose here.

*Figure 13.15 (2)* shows how we can project the sensors. *Ds* is how far the distance sensor goes out to the side (`dist_side_mm` in the code). *Df* is how far forward the sensors are from the robot wheels (`dist_forward_mm`). *Dr* is the distance sensed, from *Df*. We will have to add this to every pose. Pre-calculating a triangle from the sensed distance makes this a little easier. $\theta$ is the robot's heading. Using the *SOHCAHTOA* mnemonic, we can get $\theta r$, the angle from the robots heading to the right sensor, and using Pythagoras' theorem, we can get *Hr*, the hypotenuse. The adjacent side will be *Df* and *Dr*, and the opposite *Ds*. We can then add *Hr* at the $\theta r$ angle to each pose to get the right sensor beam endpoint for every pose. The same can be applied to the left sensor reading. While complicated, this is faster than calculating the endpoint projecting out to the side and forward for each pose.

Measure the position of the distance sensors on your robot relative to the middle of the wheels (or use the CAD drawings). The active part of each sensor to measure to is the shiny part at the top middle of each sensor.

We'll add these distance sensor positions to our `robot/robot.py` measurement after the wheelbase_mm definition:

```
dist_side_mm = 37
dist_forward_mm = 66
```

We need to add a function to `robot/arena.py` to look up a position in the distance grid:

1. The function converts an $(x, y)$ pose argument to grid coordinates:

   ```
   def get_distance_likelihood_at(x, y):
       """Return the distance grid value at the given
   point."""
       grid_x = int(x // grid_cell_size + overscan)
       grid_y = int(y // grid_cell_size + overscan)
   ```

2. Out-of-bounds requests should return an incredibly low probability:

   ```
   if grid_x < 0 or grid_x >= distance_grid.shape[0] or
   grid_y < 0 or grid_y >= distance_grid.shape[1]:
       return low_probability
   ```

3. We can then return the result stored at the grid location:

   ```
   return distance_grid[grid_x, grid_y]
   ```

In the `robot/code.py` file, we can add a method to the `Simulation` class to perform the preceding triangle calculations, to get the sensor endpoints for each pose:

4. This method will take the sensor reading, and inform us whether it's on the right side:

   ```
   def get_sensor_endpoints(self, sensor_reading,
   right=False):
   ```

5. We then calculate the adjacent and angle of our triangle:

   ```
   adjacent = sensor_reading + robot.dist_forward_mm
   angle = np.atan(robot.dist_side_mm / adjacent)
   if right:
       angle = - angle
   ```

Note here that we calculate the negative of the angle if it is on the right side.

6. We then get the hypotenuse – this will be the distance from between the wheels to the sensor beam endpoints:

```
hypotenuse = np.sqrt(robot.dist_side_mm**2 +
adjacent**2)
```

7. Now, we use numpy to help us calculate the angle relative to each pose, converting the pose angle to radians as we go:

```
pose_angles = np.radians(self.poses[:,2]) + angle
```

8. We can then build a list of endpoints by projecting from the coordinate of each pose, with the hypotenuse at the calculated angle:

```
sensor_endpoints = np.zeros((self.poses.shape[0],
2), dtype=np.float)
sensor_endpoints[:,0] = self.poses[:,0] +
hypotenuse * np.cos(pose_angles)
sensor_endpoints[:,1] = self.poses[:,1] +
hypotenuse * np.sin(pose_angles)
return sensor_endpoints
```

We finally return these calculated lists.

We now create an observe_the_distance_sensors method inside the Simulation class and apply those to the existing set of weights:

1. Start by accepting an existing list of weights as an argument:

```
def observe_distance_sensors(self, weights):
```

2. We then call get_sensor_endpoints for each side:

```
left_sensor = self.get_sensor_endpoints(self.
distance_sensors.left)
right_sensor = self.get_sensor_endpoints(self.
distance_sensors.right, True)
```

We now have a list of distance sensor projections for every pose. We can look up the distance likelihood grid at each of those points and add them:

```
for index in range(self.poses.shape[0]):
        sensor_weight = arena. get_distance_
likelihood_at(left_sensor[index,0], left_sensor[index,1])
        sensor_weight += arena. get_distance_
likelihood_at(right_sensor[index,0], right_
sensor[index,1])
```

3.  We can then multiply this by the existing weight (inside or outside the arena):

```
weights[index] *= sensor_weight
```

4.  Now, we leave this loop and return the modified weights:

```
return weights
```

We then need to incorporate this code into `observation_model`. Make the highlighted change:

```
def observation_model(self):
    weights = np.ones(self.poses.shape[0], dtype=np.float)
    for index, pose in enumerate(self.poses):
        if not arena.contains(pose[0], pose[1]):
            weights[index] = arena.low_probability
    weights = self.observe_distance_sensors(weights)
    weights = weights / np.sum(weights)
    return weights
```

If you send this code to the robot, it will weigh and resample from two distance sensors. The robot poses will start to form in blobs, located around likely positions for the sensors. The blobs will form, scatter, and reform as the robot filters and moves them. The following diagram shows what you will see on the display:

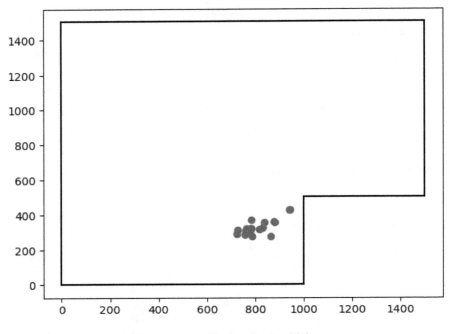

Figure 13.16 – The localization blob

In *Figure 13.16*, the poses have grouped together in a blob, roughly representing the robot's position, which will move with the robot as it drives around the arena.

You may see one blob, or a few, and they may drive exactly with the robot or seem a little off. This is where you will need to tune the model to better suit the situation.

## Tuning and improving the Monte Carlo model

Tuning the following factors can improve this model:

- The `ch-13/4.3-monte-carlo_perf` folder in the GitHub repository contains an instrumented version of the code in this chapter for troubleshooting. You will need to tune `robot.py` and `arena.py` for your own setup, but this code reports issues and tracebacks back to the computer for diagnosis, weight output from the observation model rendered on the display, and, if you are connected via USB, also sends performance data.

- Measurements in `robot/robot.py` – the accuracy of measurements such as wheel diameters, wheelbase, gear ratio, and encoders will guide the odometry model. If the movement of the blob doesn't match the speed and turning, these are the likely suspects. The model assumes wheels to be identical in size, which may be false if it's consistently pulling to one side.

- Similarly, if the distance sensor position measurements in `robot/robot.py` are incorrect or the sensors need calibration, the position will be consistently off. The sensors can be disrupted by strong sunlight and flickering room lighting.

- In the arena, the `get_distance_likelihood` factor, 100, adjusts the decay for the likelihood field around each boundary. Lowering this will tighten the fields.

- In `code.py`, the number of poses is a major factor. Increasing this will make for a better model, but beware of memory constraints on Raspberry Pi Pico.

- The `code.py` alpha factors encode your certainty in the motion model; make these lower if you trust the motion model more.

- The model also assumes the arena construction to be fairly accurate. Errors in these assumptions can stack up to make it harder for the localization algorithm.

You will need to spend time with these factors to make a model that is more likely to find its location, or is quicker in doing so.

Let's summarize what we have learned in this chapter.

## Summary

In this chapter, we started by building a test arena for our robot using foam board construction, then modeled this in code, and displayed it along with a distance sensor likelihood field. We put this on the robot, sent it over BLE, and then added poses.

We modeled how poses move using sensors, adding uncertainty to the model. We then added a model of distance sensor observations, generating weights that we used in a resampling algorithm to generate new poses.

We finished with a look at tuning factors to improve the performance of this system.

In the next chapter, we will summarize your Raspberry Pi Pico robotics learning journey so far and discuss how you can continue your journey by improving this robot or building more robots.

## Exercises

The following exercises will deepen your understanding of the topics discussed in this chapter and make the robot code better:

- The IMU could be added by storing a previous state and calculating the delta. You could mix this into the `rot1/rot2` values by taking the average of encoder calculations versus the IMU angles, or consider whether one sensor is more trusted than the others. You will need to calibrate the IMU before it can be used.

- The robot's pose guesses get stuck in **local maxima** – good but wrong guesses that are likely based on sensor positions. Consider throwing in 10 fresh guesses at every population to nudge the code to try other options.

- We are using only two observations per pose – having more distance sensors could improve this but will make the model slower.

- Could you add a target zone to the arena? Consider how PIDs could be used to steer the robot toward this. Perhaps feed the PID with the mean pose.

- You can improve the visualization by sending more pose data, including orientation. You may need to consider the `msgpack` library or switching to Wi-Fi or BLE over SPI, as the amount of data can easily overwhelm the BLE UART connection.

## Further reading

These aids for further study will let you read on and dive deeper into the Monte Carlo algorithm and its quirks:

- *Probabilistic Robotics* by Sebastian Thrun, Wolfram Burgard, and Dieter Fox, published by MIT Press, covers the Monte Carlo particle filter, along with the Kalman filter and other probability-based models in far more depth.

- I strongly recommend the *Khan Academy* material on modeling data distributions for learning and practicing data distributions.

- A playlist of 21 videos from Bonn University and Cyrill Stachniss at `https://www.youtube.com/playlist?list=PLgnQpQtFTOGQEn33QDVGJpiZLi-S1L7vA` covers the topics used here in detail. I recommend them if you want to dive far deeper into this topic.

# 14
# Continuing Your Journey – Your Next Robot

Throughout this book, you've learned how to plan, design, build, and program a robot. We've covered many fundamental topics with some hands-on experience, examples demonstrating the basics, and ideas for improving them. In this chapter, we will briefly recap our knowledge to take it further.

Thinking about your next robot, we'll answer questions such as the following – how would you plan and design it? What skills might you need to research and experiment with? What would you build?

In this chapter, we will cover the following main topics:

- A summary of what you have learned in this book
- Planning to extend this robot
- Planning your next robot
- Further suggested areas to learn about

## Technical requirements

For this chapter, you will require the following:

- Diagramming tools such as a pen or pencil and paper to sketch ideas
- Cardboard, a ruler, and cutting tools to make a test fit
- A computer with internet access
- Sketching tools such as https://app.diagrams.net/

# A summary of what you have learned in this book

As a robot builder, you have started from the basic plan for a robot. You learned about skills useful for robotics, designing, building with tools, programming, interfacing electronics, or integrating systems combining all of these, so let's dive deeper.

## Basic robotics with Raspberry Pi Pico

From the first chapter, we learned about Raspberry Pi Pico, why it's a great controller for robots, and how it stacks against others, considering the trade-offs between Pico and its larger Raspberry Pi family.

We covered ways in which Raspberry Pi Pico can be programmed and chose CircuitPython for it. CircuitPython has excellent access to hardware, with a growing library of support for many electronics modules used in robotics.

We took a tour of concepts such as the interface ports:

- GPIO to control or interface externally from Pico.
- UART, SPI, and I2C form data buses to send and receive data from devices.
- The unique **Programmable Input/Output (PIO)** peripheral lets you adapt or build interfaces.

We then planned a robot around Raspberry Pi Pico, considering its size, type, and complexity. We calculated the power requirements and specified a driver board to control motor power. We thought about the sensors we might want to use and how all these decisions are trade-offs.

We considered the GPIO pin usage, ensuring that all the chosen devices would be usable on Raspberry Pi Pico together.

We then used simple cardboard templating to test-fit our basic robot design to see whether it was a viable plan to take to the next stage.

We looked at places to shop for parts, drew up a parts list to shop for, and bought the components and tools needed to build the moving robot.

In *Chapter 3* and *Chapter 4*, after designing a robot using FreeCAD in 3D, we then built the robot using sheet plastic and simple tools.

After building the robot platform, in *Chapter 5*, we used GPIO to control the motors, demonstrating the initial movements of the robot. Then, we used this to drive along a planned path, demonstrating that we can control it in sequence and observing some of the shortcomings of control without sensors.

After this, we looked at the initial sensors on the robot.

## Extending a Raspberry Pi Pico robot with sensors

We were able to add sensors and learned how to make use of them. In *Chapter 6*, we used encoder sensors that were part of the motor assembly, and then in *Chapter 7*, we specified additional sensors – where we would fit them, the modifications we'd need to make to the chassis, and which sensors to buy. Adding sensors is where block diagrams of the robot's hardware came in handy to see what was connected where.

The following diagram shows the sensors we used in our robot:

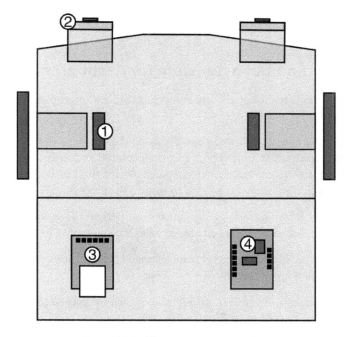

Figure 14.1 – The sensors on our robot

The preceding diagram shows the sensors we used on the Raspberry Pi Pico robot. The sensors are numbered:

1. The encoders can measure the movement of the robot's motors and, by extension, the robot's wheels. We can use these to measure distance or speed. We covered using Raspberry Pi Pico's PIO system to read these sensors continuously so that our code would not miss encoder steps.

2. The distance sensors (from *Chapter 8*) can detect how far away objects in front of the robot are. We interfaced these using I2C and looked at how to mount them and wire them in and at the programs used to fetch distance information from them. We learned how these sensors work by bouncing a signal off objects to sense them.

3. Bluetooth (from *Chapter 9*) is not a sensor but more of a communication system. We were able to build a shelf to fit the Bluetooth LE module, wire it in, and write code so we could communicate between Raspberry Pi Pico and a smartphone. We could use the phone to control the robot and display or plot data from the robot.

4. We added an **Inertial Measurement Unit (IMU)** in *Chapter 12* and learned how to use it to sense the robot's orientation by combining results from an accelerometer, gyroscope, and magnetometer. First, we looked at how to connect this with I2C and then how to calibrate (set up and orient) the sensor and get initial readings.

As well as adding the sensors, we began exploring what our robot could do with them and wrote behaviors utilizing them.

## Writing CircuitPython behavior code for Raspberry Pi Pico

We wrote programs that made the robot use sensor inputs to drive its motors in smart ways, defining these kinds of programs as behaviors.

The first behaviors (from *Chapter 5*) followed a pre-planned path with motors only; however, without sensors, this wasn't very accurate. In *Chapter 6*, we learned about the encoder sensor and measuring encoder counts, and then saw how to use them to stop driving at a fixed count. This code introduced sensor loops, with feedback from the sensors used to control the motors.

In the next section, starting from *Chapter 7*, we specified more peripherals to add. We started with the distance sensor in *Chapter 8*, where we learned how to make a behavior to avoid colliding with walls.

In *Chapter 9*, we linked up the Bluetooth device, which opened more exciting and complex behaviors by letting us observe and plot sensor data.

The scope of sensor feedback loops became more interesting when we learned about the **Proportional-Integral-Derivative (PID)** algorithm in *Chapter 10*, letting us set up smooth motor responses to stimuli. We demonstrated this with distance sensors to keep at a known distance from an object. We then used the same technique to follow walls, which was most satisfying to test with a few boxes in the middle of the room, watching the robot navigate autonomously around them.

The PID algorithm prompted us to revisit the encoders in *Chapter 11*. We converted counts into standard speed units and then used the PID algorithm to control the motor and wheel speeds to meet a value in meters per second. This refinement allowed the robot to travel in a straight line, and controlled its motion to drive it for a known distance at a desired speed; for example, we could instruct a robot to travel 1 m in a straight line at 0.17 m/s.

With the introduction of the IMU in *Chapter 12*, once we had connected and calibrated it, we learned how to use the IMU to get the robot's orientation. We then connected that with a PID algorithm to make the robot turn to face north, regardless of its initial heading.

We also looked at how to use the IMU to make a precise turn – 90 degrees or otherwise.

In *Chapter 13*, we learned about Monte Carlo simulation, with which we simulated many potential poses for the robot, using the encoders to track movements. We used sensor input to give each pose a likelihood weight. The weights were used in a resample algorithm to select the most likely poses. A combination of sensor fusion and statistics makes a robot feel smart as it estimates its position in an arena. We also demonstrated two behaviors running together, with collision avoidance running alongside Monte Carlo simulation.

This robot has some interesting capabilities, but where do we take them next?

# Planning to extend this robot

I rarely view robot projects as complete, especially those that are learning and development platforms. There are always new sensors to try, new programming algorithms to make, or simply bugs and quirks to iron out. On the other hand, there are ways to make robots more robust and cope with rougher environments, making the chassis lighter or the electronics simpler, so let's start by considering some ideas for this robot and hopefully inspire some of your own.

## Sensors you could add

The first exciting way to extend this robot is to add more sensors. Sensors are fun to program to get data from. This means you may have to figure out how to incorporate them into existing behaviors and mount them, though.

The following figure shows a selection of sensor extensions that could be interesting:

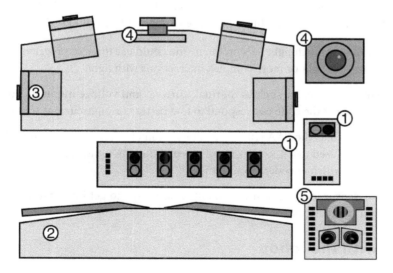

Figure 14.2 – Robot sensor extensions

The preceding figure shows a selection of sensors that would be great to extend this robot and the things you could get it to do. The sensor types are numbered:

1. **Line detectors** or **reflectance sensors**: These light up with an **Infrared** (**IR**) LED and then detect how much light reflects from an object that they point toward, or how bright/dark it is. Makers mount these under the robot as a line-following sensor. Some sensors, such as the *SunFounder 5-channel line sensor*, come with arrays of light detectors, which can be used with code to follow a line drawn on a floor. An example of a single sensor type is a TCRT5000.

2. **Bump switches**: Makers use these less in modern robots, and distance sensors should mean they aren't needed. However, you could use them so that if a robot surpasses a safe distance and bumps into something, it will immediately disengage or move back. They tend to be simple on-off switches with long lever arms to extend the edge along which they sense.

3. **Additional distance sensors**: We've used two sensors and moved them around. A set of four sensors would allow the different existing behaviors to be enhanced. It also offers more information for Monte Carlo simulation and could be used for a maze-following behavior.

4. **A camera**: There are camera sensors that we could use with Pico, such as OV7670 camera modules. They are complicated to connect and may require an additional Raspberry Pi Pico. Using downscaling and running edge detection or ML algorithms on it, it is possible to match objects. A good alternative is cameras with onboard processing, such as the HuskyLens (`https://bit.ly/3Dzurrb`). Another type of camera is a FLIR IR heat camera.

5. **LIDAR sensors**: These scan and return the depths of objects in their field of view. Having many distance sensors could extend the accuracy of Monte Carlo simulation. However, these produce a lot of data quickly and may need a more powerful CPU to control them. Solid-state sensors are low-power, small, and cheap. Pictured in *Figure 14.2* is an LDRobot LD-07 solid-state LIDAR.

Light sensors can read how much light falls on them. You could use these to program behaviors that move toward or away from light or more complex interactions with light.

A robot can have internal sensors, such as thermal, current, and voltage monitors, to monitor its batteries and motors so that the code can respond to low-battery or high-current scenarios.

Optical flow sensors, such as the PAA5100JE, facilitate odometry based on the flow of the ground below a robot and can be used to detect the overall speed of the robot; this can compensate for issues such as wheels slipping, which encoders would miss.

We've covered some sensor ideas for our robot, but another way to get information to and from a robot is through user interactions.

## Interacting with the robot

The robot in this book doesn't have many options for user interaction:

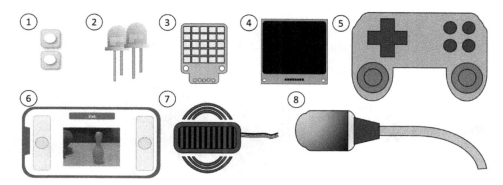

Figure 14.3 – Human interaction I/O

*Figure 14.3* shows many ways to improve human interaction with the robot:

1.  **Buttons**: Add buttons for starting and stopping a behavior on a robot.

2.  **LED lights**: Some LED lights, perhaps with different colors, can be good to show a little feedback on what is running. Big LEDs can also be used as headlights.

3.  **RGB LED displays**: These come in strips or panels, and each LED can be set to a different color. They can provide more debugging and can make cute eyes or faces too. They use multiple interfaces, a custom one-pin system, and SPI or I2C. The Pimoroni PIM435 pictured uses I2C.

4.  **OLED screens**: These can show pictures, dials, menus, text, or graphics right on the robot. These come in mono or color varieties and are frequently I2C-controlled. One such mono I2C device is the Velleman WPI438 I2C screen.

5.  **Game joypads**: A game joypad controller would be a nice way to control a robot. However, it may require a more advanced Bluetooth setup to interface with Raspberry Pi Pico.

6.  **A phone web app**: We've been using the BlueFruit app, but by swapping Bluetooth for Wi-Fi (such as with Raspberry Pi Pico-W), you can write far more interactive phone control. This will require writing more code for a frontend. At the time of writing, a system for graphing on a smartphone requires a fair amount of code to produce over Wi-Fi, and a convenient app, such as BlueFruit LE Connect, doesn't yet exist.

7.  **A beeper**: These can make beeping and buzzing sounds for the robot. You can drive some of these directly from GPIO pins, with musical tones indicating the program's state or making interesting sounds.

8.  **A microphone**: There are UART-capable voice control modules suitable for Raspberry Pi Pico. They have a small set of commands to start behaviors, and, with LED or beeper feedback, could be a novel way to interact with the robot.

User interactions can manipulate behaviors, but we need somewhere good to mount these LEDs. What can we do to the robot's chassis to improve and extend it?

## Chassis and form enhancements

The robot's chassis is simple and light, but we could take this much further, making it a more robust or aesthetically pleasing robot:

Figure 14.4 – Chassis enhancements

*Figure 14.4* has examples of two ways to improve the robot or its sensors with an outer shell:

1.  *Figure 14.4 (1)* shows a lunchbox adapted into a robot. A lunchbox, ice cream tub, or roller paint tray with holes cut into it can be made into the fairing for the robot, enclosing the electronics and holding its wiring and electronics.

2.  *Figure 14.4 (2)* shows a robot with sensor mounts that have been 3D-printed. Custom fairings and brackets can be made this way or through vacuum forming or laser cutting. See the *Design and manufacturing* section later in this chapter.

Whichever method you choose, it needs to be lightweight and easy to remove so that the electronics are accessible and you can change the batteries. It could be held in place by bolts or Velcro.

You'll need to consider how to make controls accessible – such as putting switches and charging ports on the outside. Having a fairing invites more decoration, the use of color, and style paneling. Combining this with the RGB LEDs could make for an interesting-looking robot. You could take style cues from your favorite sci-fi robot and use a similar color scheme.

## Electronics enhancements

The electronics we've used so far have been composed of modules tethered together with connector cables, using breadboards to assemble them. Breadboards are great for prototyping but take up a lot of space and weight and are susceptible to movement and vibration, with wires quickly coming loose or forming poor connections. They also look messy. What could we research to upgrade this?

Here are some sturdiness enhancements we can make to the robot electronics:

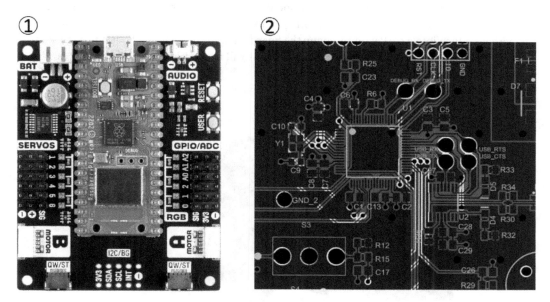

Figure 14.5 – Improving robot electronics

Electronics in the robot could be made smaller, lighter, and tidier. Try these ideas:

1.  Pictured in *Figure 14.5 (1)* is an Inventor 2040 W from Pimoroni (`https://shop.pimoroni.com/products/inventor-2040-w`). Motor and robot boards for Raspberry Pi Pico or incorporating an RP2040 will reduce the amount of wiring. The Inventor 2040 W includes motor control, lights, and servo connections.

2.  Soldered boards will be tidier. An intermediate option is using a stripboard or perfboard. These pre-made printed circuit boards can mimic the wiring arrangement of breadboard strips, allowing you to transfer breadboard designs onto them. You can then solder parts and wires into them. Wires going to external parts, such as motors and sensors, will have their connectors soldered into the board.

    *Figure 14.5 (2)* shows a custom **Printed Circuit Board** (**PCB**) from `https://github.com/uwrobotics/MarsRoverHardware`. You can download or design your own using software such as KiCad. This is complicated but gives you lots of options for customization. You use a PCB view to lay out a circuit board and route connections between parts. Using PCBs allows you to make small, light, and tidy robot designs. In addition, it opens your designs to using surface mount electronics, which makes more components available. You can use **design rules** to ensure that the connections are all made and tracks are not touching. After this, you can then prepare the part for manufacture. You can take these designs to a board house, such as Seeed Studio (`https://www.seeedstudio.com/`), that prints the board for you.

Some board houses offer a **Printed Circuit Board with Assembly (PCBA)** service, where, at additional cost, they will solder on components, sockets, and connectors before sending the part to you. Using a PCBA service means you just need to plug in the external sensors and perhaps a Pico into a socket on receiving the board. That means that you do not need to solder surface mount parts yourself. Just be aware that this will take experience, and you may produce boards that aren't right the first time.

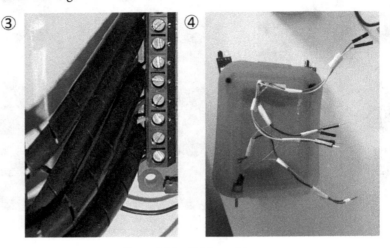

Figure 14.6 – Tidying wiring

3.  You can tidy robot cabling using 6 mm spiral wrap and cable clips, as shown in *Figure 14.6 (3)*, or just masking tape, as shown in *Figure 14.6 (4)*. Wiring can be aesthetically pleasing if you make an effort to route it nicely.

4.  For connectors, we have used Dupont connectors due to their convenience. However, once you are on a soldered circuit board, you can consider locking connectors, such as JST or Molex PicoBlade connectors. These lock cables in place so that they will not vibrate loose or easily be pulled out, and add further protection by being polarized. You will need to get practice with a crimping tool to use these effectively, but it will be worth it for better robot builds.

With a robust chassis and tidier electronics, perhaps you can get more ambitious with further outputs. Let's see things you could make the robot's outputs do.

## Outputs you could add

Outputs mean a robot could do more to move or alter its surroundings. Additional motors, such as servo motors, can be used to make interesting mechanical devices, for example, those shown in the following figure:

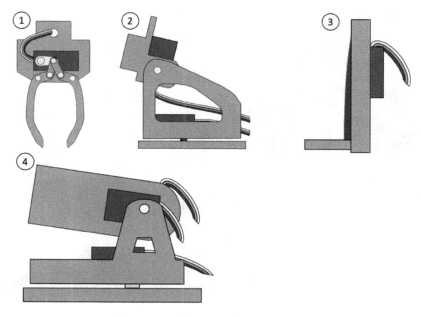

Figure 14.7 – Motors and outputs in use

*Figure 14.7* shows a few interesting ways to use further motors to extend the robot:

1.  **A gripper** – This could grasp items and move them around. The Pololu Micro Gripper Kit uses a single servo to open or close the jaws. These are fun to attach to the front of a robot.

2.  **A pan-and-tilt mechanism** – Putting sensors on this means you can direct a sensor at a particular point of interest or use the head to sweep with a sensor. One servo motor rotates the head left and right (panning), and the other tilts the head up and down. The Adafruit Mini Pan-Tilt Kit is a good example of this.

3.  **A lift-and-lower mechanism** – These use a motor to lift and lower a platform, like a forklift. This could be combined with the gripper to make a robot that can stack objects.

4.  **A ball launcher** – A launcher would be fun so that a robot could aim at targets. Robotics competitions such as PiWars have events that involve using these. Motors are required to direct the aim, and a motor or actuator is required for launching the ball.

These are just a few examples; there are plenty more mechanisms you can either buy or build. You could even add a whole arm with a kit such as the MeArm Servo Robot Arm Kit.

Now that we've seen some outputs, perhaps we can better use these sensors and outputs with more behaviors.

## Extending the code and behaviors

The robot's code and behaviors are what bring the robot to life. However, there are many ways we could improve the code on this robot, as shown in the following figure:

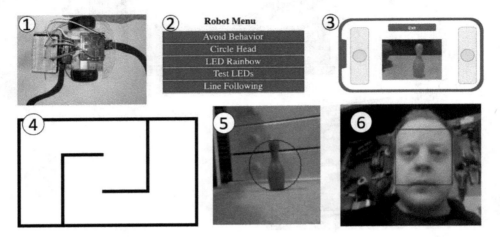

Figure 14.8 – Behavior suggestions

*Figure 14.8* depicts some suggestions; the following list covers these and more:

1.  **Line-following**: Using the line detection sensors previously mentioned in the section on sensors, you could program a robot to follow lines drawn on the floor. Line-following can use *if-then*, *bang-bang* control, or a PID algorithm for smooth line-following.

2.  **A menu system**: You can combine all the behaviors in a menu for selection. Consider how each program would be tidied up so another can take over. A variation of this is using a voice control module, along with beeps, to indicate the current robot mode.

3.  **Driving with a camera**: Adding a camera and serving it to a phone web app means you could drive the robot with a camera view – a robot periscope.

4.  **Maze solving**: With more distance sensors, the robot would be able to look for openings and find its way around a maze. Depending on the complexity of the maze, simple rules such as always turning left might work, or the robot might have a map of the maze and use the Monte Carlo method with precision navigation. For example, a turning-left method using the encoders might be able to memorize what turns it took and where, backtrack, and try other routes.

5. **Camera navigation**: Camera images can be scaled to low resolutions, and algorithms can be used to pick out features where image intensity or colors have changed over a threshold. The detected feature locations can then be combined with a PID algorithm to orient the robot relative to such features.

6. **Camera recognition**: Recognition is considerably more complex and may require looking at a machine learning system such as TinyML. See `https://bit.ly/3xy3twx` for news on TinyML ports for CircuitPython. You can use similar techniques with microphones for speech recognition, but this may be intensive enough that multiple Raspberry Pi Pico boards would be required.

The following suggestions are not pictured but represent advanced behaviors:

- **Simultaneous Location and Mapping** (**SLAM**): A robot can use its sensors to build up a map of its location and then keep track of its pose relative to what it has already mapped. This technique would use several sensors together and benefit from the LIDAR sensors that we have already mentioned. However, this advanced technique can be a deep rabbit hole!

- **Task planning** and **motion planning**: Combining controlled turns, controlled movements, and location tracking with a hopper or gripper would mean you could make behaviors to seek out, collect items, and place them in a collection point.

These suggestions should hopefully inspire some creative behaviors. There are endless possibilities for combinations of code, sensors, and outputs, which you can use to improve your knowledge, solve problems with a robot, and extend your toolbox.

So far, though, these suggestions have been focused on extending the existing platform. So, what happens when we extend to thinking about your next robot platform?

# Planning your next robot

In *Chapter 1*, *Planning a Robot with Raspberry Pi Pico*, we saw a few styles under the *What style of robot is suitable?* section. Now, inviting a greater level of imagination, let's revisit how different a robot can be.

## Form, shape, and chassis

We'll start by considering the styles from *Chapter 1* again and how you might get there:

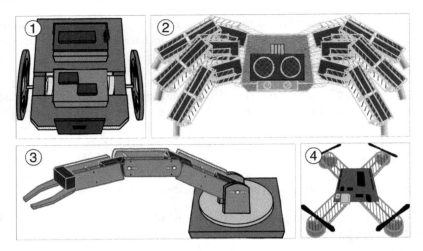

Figure 14.9 – Different robot styles

The preceding figure shows four styles of robots. We can now imagine more variations on these:

1. **A larger wheeled platform** to hold more sensors and electronics, perhaps with a more interesting wheel arrangement.

2. **A hexapod robot** with six legs – for exploring walking and gaits. This will use lots of servo motors.

3. **A robot arm**, such as the MeArm mentioned previously. These can be servo motor-based or stepper motor-based.

4. **A quadcopter drone** is super-light, but involves more interesting IMU and PID use.

The following diagram shows a few other robot ideas:

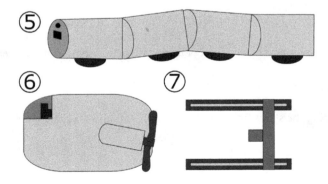

Figure 14.10 – More robot types

*Figure 14.10* shows a few more types of robots you could consider for your next project; these are potentially more ambitious as far as the ideas go:

5.  A **snake robot** is made of multiple robot segments with distributed electronics. The segments can flex like carriages on a train, with power and control running along between the segments. You can use servo motors to control the flex, and motors on segments driving wheels or tracks provide traction to pull the snake along with sensors. The front segment usually carries forward-facing sensors.

6.  A **submarine robot** would be fun to use to explore lakes or ponds. These require thinking about how you would power an immersed motor, how you might safely retrieve a robot if it loses power, and how you would control it through water, which can block RF signals such as Bluetooth or Wi-Fi. A tether cord may be necessary. You'll need to think about weighting, ballasts, and movable thrusters. A fun variation may be an amphibious floating robot that can drive on land, move in the water, and dive.

7.  An **XY** or **gantry-type robot** frequently uses stepper motors, arranged so one set moves along each axis. 3D printers, CNC mills, and laser cutters frequently use this type. They can also be used to plot images or with a grabber to pick up objects and place them somewhere.

These robot variations should inspire you. Next, let's look at a few variations on the wheeled platform.

## Variations on wheels

The wheeled platform is still a practical and straightforward robot style. However, you can extend those wheels in many exciting ways:

- **Caterpillar tracks** can offer more traction than wheels and deal with uneven surfaces. They also look attractive. Some types use a single rubber molded track, such as the tracks on the tiny Zumo chassis, and others use tracks made of links, such as those on the Devastator robot chassis. Depending on these types, you may run into problems with friction when trying to turn, and may need more torque depending on the surface.

- **Mecanum wheels** are unique. As well as moving forward/back and turning, mecanum platforms can also "*crab-walk*" sideways and drive in almost any direction; you will need four independent motors and some specialized control algorithms for this implementation. See `https://bit. ly/3y1Kjzp` for an example.

- **Tristar wheels** are clusters of three wheels set up so that either individual wheels can turn or whole clusters can turn; this allows them to climb up steps depending on their relative size. They are mechanically complicated, but you can drive them like regular wheels. They should be a four- or six-wheel drive combination.

- **Independent drive** uses individually steerable wheel pods; the Mars Discovery rover has these, with a stepper or servo motor turning each wheel assembly, and each wheel being individually drivable too; they can both crab-walk and steer. These are more complicated than mecanum wheels but are robust.

Now, we have some idea of the next robot's form and ways we could build on the simple wheeled chassis. What process would you use to design the robot?

### Design stages

You would design a robot such as this by starting with sketches and simple cardboard parts as we did in *Chapter 1, Planning a Robot with Raspberry Pi Pico*, and then take the design to CAD as you become clearer on what it is you want.

We will look at some of the electronics further, but you'll want to incorporate those into the CAD designs so you are clear on the dimensions and where to attach things before fabricating any parts for a robot.

Let's look at the electronics.

## Electronics and sensors

You will need to adapt the electronics for each form and use case. For example, larger robots with larger motors will naturally need larger motor controllers and power systems capable of handling more significant power requirements. Larger motor drivers include the LM2575, capable of handling 15 A motors at 36 V.

Robots with servo motors will need controllers such as the Adafruit 16 servo controller to handle the power and control of many servo motors. Robots with stepper motors may need similar breakouts for stepper motors, although depending on the precision needed, a DC motor controller with four channels can also drive stepper motors.

Submarine robots need to be waterproof, and quadcopters need high efficiency. For these, brushless motors are most suitable, and they will need to be driven by **Electronic Speed Controllers (ESCs)**. Some wheeled robots also use brushless motors for their great efficiency.

An integrated motor controller such as the Pimoroni Inventor 2040 W (seen in the *Electronics enhancements* section) or the Pimoroni Servo 2040 might make sense for smaller robots. For example, the Servo 2040 can control 18 motors, making it ideal for a 6-legged hexapod robot.

Where more battery power is required, you may need an upgrade from AA batteries to Li-ion or LiPo technology. In addition, you will need to carefully consider battery management systems that prevent catastrophic events and ensure you have chargers for them. The Lipo SHIM from Pimoroni will help charge these and power Raspberry Pi Pico.

In terms of sensors, you can consider the full range described previously. When planning them, consider which Raspberry Pi Pico pins will be in use. `https://pico.pinout.xyz/` is an excellent resource for this purpose.

As you use more pins and sensors, you may have to use more I2C sensors, which may require I2C multiplexers, such as TCA9548A modules. If the sensors are not I2C and simply need GPIO pins, then a PCF8574 adds other GPIO pins over I2C. If you need to add more complex sensors, a second Raspberry Pi Pico or RP2040-based board might be necessary.

You can also consider adding I2S audio amplifiers for speakers for sounds more interesting than beeps and microphones for sound recognition on Pico.

It is also worth considering whether the robot application requires a powerful CPU such as a full Raspberry Pi, and reserving Raspberry Pi Pico as an IO coprocessor.

For a robot such as a snake robot, you may consider how you'd need to wire the modules throughout the segments. If you build a submarine robot, how will you protect electronics against water getting in?

Next, we will look at the kind of code you might try with these designs.

## Code and behavior

Choosing the code has a few factors:

- What control code do you need for the sensors, outputs, and mechanisms?

- How smart does the robot need to be to solve specific problems?

- What safety factors might be needed, and how will this interact with the form and shape? For example, what control system might you need to stop a robot quickly?

The control mechanism for a legged robot may need code to group servo motors into legs with multiple joints and then smoothly move between positions defined by gaits – walking strategies. These are generally sequenced patterns; they may adjust or carry on following a sequence depending on feedback from the legs or foot sensors. Usually, the gait code controls the movement, with another program steering this gait code in a **horse-and-rider** configuration.

With a robot arm or snake robot, researching **inverse kinematic** algorithms would help to position parts of the robot relative to other parts, choosing the angles at which the servos along the arm or the snake need to be to reach this point.

Robots that use a more powerful CPU running Linux may justify going beyond simple Python scripts into **Robot Operating System (ROS)**.

If speed is an issue, you can consider a fast controller such as a Teensy, a powerful controller such as Raspberry Pi, or explore other programming languages such as C and Rust. You can use C to extend CircuitPython and Python to glue together different robot functions.

If you start to need multiple RP2040 controllers, how will they communicate and interact with each other? Via an I2C bus or UART? You may need to research and consider a protocol for them to send requests to each other.

You've now considered the robot's form and chassis, the electronics it might need, and the code you will want. Learning about more ways to build robots or situations in which to test and demonstrate code can help you build these new robot designs.

# Further suggested areas to learn about

You have some ideas for a future robot (or even a few future robots) that you want to build. Learning about some other skills will allow you to take those designs further and get more creative. Let's dive in.

## Electronics

Refer to the *Electronics enhancements* section earlier in this chapter for inspiration here. To recap, see the following:

- Learn about designing circuits with stripboard or veroboard and further soldering skills.
- Learn more about designing PCBs with tools such as KiCad. You place parts such as Raspberry Pi Pico (or, if you are braver, an RP2040) into the schematic editor to work out their connections.
- Consider how to keep the cabling tidy, perhaps designing cable routing into drawings and CAD designs for a very tidy robot.
- Use connectors that reduce vibration and connect only one way to reduce mistakes.

There are further ways to extend your robots toolbox:

- Oscilloscopes to view circuit waveforms, great for looking at PWM (see the section *An introduction to pulse width modulation speed control* in *Chapter 5*)
- Logic analyzers to debug data buses and logic systems
- Bench power supplies to test electronics without worrying about batteries
- More advanced soldering stations
- Clamps to hold boards and components in place and test those connections
- A stock of electronic components, such as diodes, resistors, capacitors, and wires
- Some standard chips, such as op-amps and regulators, along with motor spares

Another advanced electronic capability is to look at **Field-Programmable Gate Arrays** (**FPGAs**). These devices allow you to program digital circuitry into them, allowing fast I/O helpers beyond PIO and even small CPU cores. They are not cheap or easy to use, but they offer huge flexibility in interfacing and prototyping new chips.

We can look at advanced manufacturing techniques now that you've seen some advanced electronics techniques.

# Design and manufacturing

We've used hand tools to build this robot. However, some techniques allow you to create far more intricate robot shapes. We touched on them in the *Chassis and form enhancements* section in this chapter.

The gateway to many of these is practicing using 3D CAD and learning about Blender to create more organic forms. It is worth experimenting with 3D CAD alternatives to FreeCAD, such as Solvespace or Fusion 360, and seeing which work for you. Finally, it is worth learning about the Inkscape program for 2D drawing or making decals to place on the 3D forms. During design, you should also be clear on how to assemble the parts when they arrive.

All fabrication methods require considering the limitations of the systems at design: for example, minimum cut widths, part thickness to avoid breaking, and how a cutting tool might access the different surfaces in a part.

Let's see a list of how you can manufacture parts:

- **Laser cutting** is a natural extension of how we've been cutting parts. The CAD output is precision shapes to be cut into a 2D sheet of material. Designs are assembled like flat-pack furniture or using stand-offs. You can make impressive designs with this fabrication system. The MeArm (`https://mearm.com/`) robot arm or OhBot robot head (`https://www.ohbot.co.uk/`) are made with laser cutting.

- **CNC milling** lets you make cuts into material with a robotically controlled cutting part, and can cut wood, plastic, and metal. It can cut out parts with different depths forming complicated shapes, but mostly operate from above and cannot make cuts from the side or below.

- **3D printing** presents the possibility of fully 3D intricate parts, especially when they may be one-off parts for a single robot design. You can iterate with these, printing refinements to parts as you improve a design. Desktop 3D printers are inexpensive and can make sense in a home lab.

- **Vacuum forming** is where a plastic sheet is pulled against a mold (or a buck) to create a shell. This technique allows for thin yet single-part plastic areas and may be perfect for robot shells (fairings). You can use the other techniques mentioned to make the buck.

- **Metal techniques** such as welding, cutting, and using sheet metal may be useful for large robots or those that might be handled roughly. However, most small hobby robots will not need this. Some metal construction can be made simply by using aluminum extrusion, a hacksaw, and t-slot bolt heads.

Where can you get these manufactured? Many of these machines are large and expensive. If you do not have the space at home, you can consider sending designs to a company and have them cut them for you, such as `https://razorlab.online/`. If you are part of a school, college, or university, it may have these devices in its labs. Otherwise, see the *Places to build robots* section later in this chapter for information about maker spaces.

The preceding techniques will let you make interesting and varied robot designs, improving on parts and exploring different shapes. Access to these techniques and people to help you learn about them becomes easier if you get involved in robotics communities.

## Robotic competitions and communities

There are many robotics communities. This section does not aim to be exhaustive but to get you started with getting involved with them.

### Social media

This book has a Discord community at `https://discord.gg/2VHYY3FkXV`. You can use this to ask me questions and discuss your robotics with other robot builders.

Adafruit has a Discord community for discussing robotics, electronics, and CircuitPython – find out more at `https://blog.adafruit.com/2019/02/05/adafruit-community-server-on-discord-now-included-in-the-open-source-listings-discordapp-discord-opensource-circuitpython/`.

There is a lively robotics community on Twitter. I am on there as `@orionrobots`, and I regularly share and boost robotics tweets; I will answer robotics questions and introduce robot builders into the community.

Many of the following communities have Twitter handles or tags, which are worth checking out. Twitter also has a `#MakersHour` tag for people talking about making anything, including robots.

I am also available on YouTube at `http://youtube.com/orionrobots`, where you can see the robots I've been building, on Mastodon at `https://fosstodon.org/@orionrobots`, and on Facebook at `https://www.facebook.com/orionrobots`.

The robot builder James Bruton showcases inspiring robot builds as `@XRobotsUk` at `https://twitter.com/XRobotsUK` and on YouTube at `https://bit.ly/3RYScxp`.

Another great robot builder with a community is Kevin McAleer at `https://www.youtube.com/c/kevinmcaleer28`. In addition, he has a lively Facebook robot community at `https://www.facebook.com/groups/smallrobots` for discussing exactly the kinds of robots we have built in this book and extending far past this.

### Events

The PiWars event is a competition held in the UK for robot builders using Raspberry Pi to compete in autonomous and manually driven challenges, showing off and improving robot building, robot driving, and coding skills. The community is global, welcoming, and happy to share techniques. You can find its website at `https://piwars.org/`. PiWars also has a `https://twitter.com/piwarsrobotics` account and a PiWars Discord server at `https://discord.gg/sjABKje`.

The **UK Micromouse and Robotics Society** (**UKMARS**) community runs robotics competitions, with Micromouse being the oldest. Micromouse is a maze-running robot challenge running in the UK, with a lively community of robot builders. Visit `https://ukmars.org/` for more information.

The **First Robotics Competition** (**FRC**) at `https://bit.ly/3BnWOWG` inspires people to build robotics and compete at robotics globally. The events occur in some local areas, with championships eventually taking place in Houston, Texas. The regional events mean that in-person competitions and collaboration can take place with teams long before traveling to Texas.

Maker faires are events held all over the world, and there may be some taking place in your country. Makers come to exhibit, talk about, and celebrate their creations. These are based around maker communities and provide great inspiration and contact between makers of all kinds, including robot builders. See `https://makerfaire.com/` for details, including a search for maker faires near you.

Where can you build these robots if you need more help, tools, or space than you can get home?

### Places to build robots

Tools and experience are important for building robots. While talking online can help, little is as useful as working with people experienced in the use of certain tools. Where can you find a space like the one in the following figure?

Figure 14.11 – A maker space

Maker spaces, hackspaces, or **Fabrication Laboratories** (**Fab Labs**) such as the one illustrated in *Figure 14.11* are places where you can come and gain access to advanced tools such as 3D printers, laser cutters, and CNC machines, along with arrays of hand tools, electronics equipment, and large workbenches for them. They are usually accessible via membership and a small donation.

These spaces are usually well lit. Knowledgeable people maintain the tools and can also help you with your builds, with advice and training for the tools. An expert may also suggest a different manufacturing technique that may get you the results you want in a better way (quicker, cheaper, or stronger).

These spaces also tend to have safety systems, such as fume or dust removal, which is essential when soldering electronics or cutting materials such as wood. You will also find that these spaces usually have components or materials to hand that are needed to try a new technique. Using a small amount of material or borrowing a tool is essential if you want to try a technique before buying a lot of equipment. Searching for a Fab Lab, maker space, or hackerspace near you is strongly recommended.

Some coder dojos offer robotics programming as part of their courses. Coder dojos will focus more on the code and algorithms and less on building the robots. Schools that run code clubs or STEM clubs may also be amenable to robot programming and building.

Now that we've seen some communities, what further areas are there to explore in the code area?

## Robotics systems and code

The software that runs on robots has a huge potential for exploration. With the ability to add additional controllers and memory or expand to larger and more powerful controllers, there are few limitations in this area.

We've seen in *Chapter 13, Determining Position Using Monte Carlo Localization*, how a robot can use simulation to help the robot understand its world, so you may even consider expanding upon that in a virtual environment too. While that is appropriate for algorithms on the robot, extending this to full visualization and 3D would be reinventing the wheel. Simulation using systems such as the Godot game engine or the ROS Gazebo system will let you start to test robot algorithms away from the actual robot, letting you improve planning and SLAM techniques using sensor data. You can build 3D, physics-enabled worlds for a robot and test code there. There are considerations about being able to transfer code between the languages used in the simulations and CircuitPython that you'll need to resolve. However, you may consider languages other than CircuitPython as you move to larger projects. You may be able to tune some things, but be aware that PID algorithm values will likely change when you attempt to try the same robot code in the real world.

Building on the idea of location mapping concepts are goal-based algorithms and task planning (mentioned in the *Extending the code and behaviors* section). These consist of planning how a robot can reach one location from another and what a robot may need to do if it has to move items to complete that operation. For example, the PiWars eco-disaster (`https://bit.ly/3xBPIwz`) challenge required planning how to move barrels to a specific location without knocking over others. With random barrel locations, this is a good challenge to use to learn about these advanced robot algorithms.

We've briefly mentioned machine learning before. There are a few different techniques, with variations on neural networks being the most popular type at the time of writing. Deep learning offers image and sound recognition results with TensorFlow on big controllers and TinyML on small ones such as Raspberry Pi Pico. These extend to image analysis, as well as image classification. Combining this with advanced computer vision techniques and navigating with a camera or two as the sensors become possible. However, LIDAR sensors are still a reliable source of distance information.

Another interesting technique is clustered robots, a group of robots working as a single system. These would need to communicate with each other using RF, such as Wi-Fi, Bluetooth, or IR. You will need code that can plan how multiple robots will solve a problem. I recommend researching bird flocking and other artificial life subjects to build up to solving this, along with the planning techniques mentioned previously.

You've now seen algorithms to extend your learning about robotics further. Let's recap what we have seen.

## Summary

In this chapter, we have gone back over what we have learned in *Raspberry Pi Pico Robotics for Workbench Wizards* and the robot we've built using these skills.

We then investigated how we could extend this robot, building additional features such as sensors, outputs, and a more interesting or robust chassis, and extending the code to do amazing things.

We branched out further, providing ideas and inspiration or areas of research for your next robot, things you could design and build with a clean slate.

Finally, we dove into advanced techniques in electronics, ways to manufacture far more intricate parts, the robot communities and spaces you could be part of, and where robotics programming can go. This last section should provide plenty of inspiration for continuing to build robots!

## Exercises

The following exercises will deepen your understanding of these topics and make the robot's code better:

- Consider an interesting improvement to the current robot, plan it, and get started.
- Join a robotics community and get involved with robots on Discord, Facebook, Twitter, or Mastodon.
- Plan and build your next robot or robotics-related gadget. Then, share it with the communities!

Your robotics journey is just starting now. I look forward to seeing you in the robotics community, along with the machines you create!

# Further reading

You can continue your robot-building journey with the following books and reading areas:

- In *Learn Robotics Programming* by *Danny Staple, Packt Publishing*, I have written about building robots with Raspberry Pi, with another wheeled robot build, visual processing, and speech control covered, along with a different power system.

- *Python Robotics Projects* by *Prof. Diwakar Vaish, Packt Publishing*, shows you how to build several small robots, explores the code in them, and has machine learning projects to try with them.

# Index

# Q

`Packt.com`

Subscribe to our online digital library for full access to over 7,000 books and videos, as well as industry leading tools to help you plan your personal development and advance your career. For more information, please visit our website.

## Why subscribe?

- Spend less time learning and more time coding with practical eBooks and Videos from over 4,000 industry professionals

- Improve your learning with Skill Plans built especially for you

- Get a free eBook or video every month

- Fully searchable for easy access to vital information

- Copy and paste, print, and bookmark content

Did you know that Packt offers eBook versions of every book published, with PDF and ePub files available? You can upgrade to the eBook version at `packt.com` and as a print book customer, you are entitled to a discount on the eBook copy. Get in touch with us at `customercare@packtpub.com` for more details.

At `www.packt.com`, you can also read a collection of free technical articles, sign up for a range of free newsletters, and receive exclusive discounts and offers on Packt books and eBooks.

# Other Books You May Enjoy

If you enjoyed this book, you may be interested in these other books by Packt:

**Raspberry Pi Pico DIY Workshop**

Sai Yamanoor, Srihari Yamanoor

ISBN: 978-1-80181-481-2

- Understand the RP2040's peripherals and apply them in the real world
- Find out about the programming languages that can be used to program the RP2040
- Delve into the applications of serial interfaces available on the Pico
- Discover add-on hardware available for the RP2040
- Explore different development board variants for the Raspberry Pi Pico
- Discover tips and tricks for seamless product development with the Pico

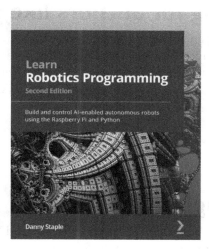

**Learn Robotics Programming - Second Edition**

Danny Staple

ISBN: 978-1-83921-880-4

- Leverage the features of the Raspberry Pi OS
- Discover how to configure a Raspberry Pi to build an AI-enabled robot
- Interface motors and sensors with a Raspberry Pi
- Code your robot to develop engaging and intelligent robot behavior
- Explore AI behavior such as speech recognition and visual processing
- Find out how you can control AI robots with a mobile phone over Wi-Fi
- Understand how to choose the right parts and assemble your robot

# Packt is searching for authors like you

If you're interested in becoming an author for Packt, please visit `authors.packtpub.com` and apply today. We have worked with thousands of developers and tech professionals, just like you, to help them share their insight with the global tech community. You can make a general application, apply for a specific hot topic that we are recruiting an author for, or submit your own idea.

# Share Your Thoughts

Now you've finished *Robotics at Home with Raspberry Pi Pico*, we'd love to hear your thoughts! Scan the QR code below to go straight to the Amazon review page for this book and share your feedback or leave a review on the site that you purchased it from.

`https://packt.link/r/1803246073`

Your review is important to us and the tech community and will help us make sure we're delivering excellent quality content.

# Download a free PDF copy of this book

Thanks for purchasing this book!

Do you like to read on the go but are unable to carry your print books everywhere? Is your eBook purchase not compatible with the device of your choice?

Don't worry, now with every Packt book you get a DRM-free PDF version of that book at no cost.

Read anywhere, any place, on any device. Search, copy, and paste code from your favorite technical books directly into your application.

The perks don't stop there, you can get exclusive access to discounts, newsletters, and great free content in your inbox daily

Follow these simple steps to get the benefits:

1. Scan the QR code or visit the link below

https://packt.link/free-ebook/9781803246079

2. Submit your proof of purchase

3. That's it! We'll send your free PDF and other benefits to your email directly